Gear Cutting Practice, Machines and Tools

by

J. C. CROCKETT
C.Eng., M.I.Mech.E., M.I.Prod.E.

THE MACHINERY PUBLISHING CO. LTD.
Head Office: NEW ENGLAND HOUSE, NEW ENGLAND STREET,
BRIGHTON, SUSSEX, BN1 4HN.
Reg. Office: CLIFTON HOUSE, 83-117 EUSTON ROAD, LONDON N.W.1.

First Published 1971

Copyright © Machinery Publishing Co. Ltd.

Author: J. C. Crockett

ISBN 0 85333 211 8

Printed in Great Britain by
The Grange Press, Southwick, Sussex

CONTENTS

SECTION 3: INSPECTION

ACKNOWLEDGEMENTS

Acknowledgements and grateful thanks are due to the following who have supplied photographs and information.

W. E. Sykes
Churchill Gear Machines
Hermann Pfauter (Germany)
Barber Colman (U.S.A.)
Fellows Gear Shaper Co. (U.S.A.)
Michigan Tool Co. (U.S.A.)
Precision Gear Machines & Tools
Hey Engineering Co.
Ernst Grob (Switzerland)
Mico-Collette (Belgium)
W. Ferd. Klingelnburg
British Standards Institution
Goulder & Son
Coventry Gauge & Tool Co.
N.E.L. East Kilbride
Drummond Bros.
Cambridge University

INTRODUCTION

Many books have been written on gears and gear geometry but few have dealt with the practical function of gear cutting. The object of this book is to deal with the problems of gear cutting and the aspects which confront the gear engineer or technician in his normal daily routine. To work efficiently he needs a knowledge of the machine tools in common use, the best type of machine for a given job, the small tools used to produce the gear, methods of sharpening the tools and inspection techniques for machine tools and gears. He must have information on the calculation of production times and the design of work holding fixtures, and finally, of course, a basic knowledge of the gear itself and how it works. Every element is virtually of equal importance; the accuracy produced by an expensive machine tool can be ruined by errors introduced by faulty sharpening of the cutting tool. Attention must therefore be paid to each element in turn if optimum results are to be obtained.

The improvement in performance of gear-cutting machines over the last three years, centred mainly in Europe, has been very impressive. Productivity and accuracy achieved during this period, particularly on certain gear-shaping machines, have been greater than can normally be expected. Much of this improvement is due to the attention paid to the kinematic drive and the stiffness of the structure, both static and dynamic.

The design and manufacture of gear-cutting machines and tools are being constantly improved and this book records the state of the art at the date of writing.

PRODUCTION METHODS AND MACHINES

PART 1.1: HOBBING

1.1.1 Principle of the hobbing process

Gear hobbing is a continuous generating process in which the tooth flanks of the constantly moving workpiece are formed by the equally spaced cutting edges of the hob. The profile produced on the gear is a curve comprising a number of flats varying with the number of flutes in the hob which pass a given tooth during the generating movement (Fig. 1).

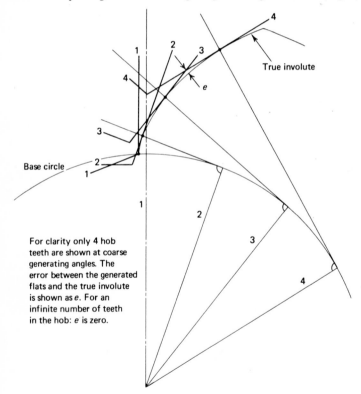

For clarity only 4 hob teeth are shown at coarse generating angles. The error between the generated flats and the true involute is shown as *e*. For an infinite number of teeth in the hob: *e* is zero.

Fig. 1. Involute generation by a hob

All generating processes have this common feature in that the profile is not a smooth continuous curve but is built up of a series of steps or flats. If the hob could be considered as having an infinite number of edges then it would produce a smooth profile. The hob is in fact a worm provided with cutting edges and suitably relieved to give cutting clearance. The pitch of the hob measured normal to the helix angle is equal to the pitch of the gear measured around its pitch circumference. For one revolution of the hob, therefore, a given tooth will engage once in each tooth space of the gear. Obviously the more flutes in the hob the more cutting edges available for forming the tooth flanks and the smoother the resulting curve.

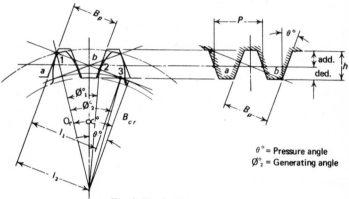

$\theta°$ = Pressure angle
$\emptyset°_2$ = Generating angle

Fig. 2. The hobbing process

Fig. 2 shows clearly the action of the hob flutes on the gear flanks. The distance B_p is the base pitch which corresponds to the pitch of the gear measured around the base circle circumference,

i.e. $B_p = P \cos \theta°$ or $B_p = \dfrac{2B_{cr} \times \pi}{N_{tg}}$ where $\theta°$ = pressure angle and N_{tg} = the number of teeth in the gear.

The number of flutes in the hob required to generate the complete tooth flank can be found as follows:

$$\text{Where } \cos \alpha° = \frac{B_{cr}}{O_r},$$

$$l_1 = \text{length of line of action,}$$

$$F = \text{number of flutes in the hob}$$

$$l_1 = \frac{\text{ded.}}{\sin \theta°} + \left(\frac{\text{add.} - O_r \text{ versine } (a° - \theta°)}{\sin \theta°} \right)$$

B_{cr} = Base circle radius
O_r = Outer radius

number of revolutions of hob to generate a flank $= \dfrac{l_1}{B_p}$

number of flutes required $= \dfrac{l_1}{B_p} \times F.$

The diagram shows clearly that generating starts with hob tooth (a) at point (1) on the gear flank and by the time the hob has made one revolution tooth (b) on the same flute will contact the gear flank at point (2). It can be seen, however, that the gear flank has not been completed – the generating angle is $\emptyset_1^\circ$ but the gear is required to rotate through $\emptyset_2^\circ$ before the generation is complete.

Advantages and limitations of the hobbing process

1 The main advantage is its versatility in that it can cover a variety of work including spur gears, helical gears, worms and wormwheels, splines and serrations and a variety of special forms.

2 All the working components of the machine are continuously rotating and there is no reversal of motions during the cutting operation as there is in the shaping process.

3 The indexing is continuous and there is no intermittent motion to give rise to errors.

4 There is no loss of time due to non-cutting on the return stroke.

5 There are no reciprocating parts and therefore none of the problems due to reversal of masses.

It is also possible to generate internal gears but the application is very limited and involves a special hob head and cutting tools; even then the gear diameter must be reasonably large. The double helical type gear can also be hobbed but only where a large gap is allowed at the centre between the two helices.

The obvious limitation is that some gears are restricted by adjacent shoulders larger than the root diameter and so close as to restrict the approach of the hob. Such gears can be produced only by the gear shaping process.

Splines and serrations are sometimes required with one tooth blocked or removed such that it can be engaged only at one position; this type of component is not suitable for hobbing but is ideal for shaping.

Special profiles or cams which are virtually one tooth components are also impractical to hob for two reasons – the difficulty of making the hob and the index ratio of 1:1 hob and work.

1.1.2 Principle of the hobbing machine

The generating process requires accurate relationship between various elements of the machine in order to produce the desired results. Change gears on many types of machine tools can be chosen independently of other functions of the machine but this is not true on a hobbing machine, particularly when hobbing helical gears, as all the motions of the machine must accurately be related to each other. It is the relationship between the rotation of the hob, the rotation of the work and the amount and direction of feed which determines the gear to be cut.

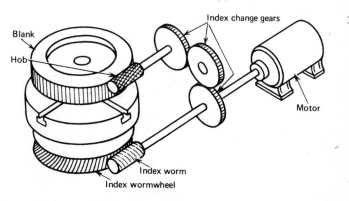

Fig. 3. The principle of the hobbing machine

The basic principle of the hobbing process is illustrated diagrammatically in Fig. 3. To reduce the kinematic train to its simplest form it has been assumed that the table index worm and the hob are of the same hand of helix and number of teeth. Under these conditions the expression for the index change gears is $\dfrac{N}{n} = \dfrac{DR}{DN}$, where N is the number of teeth in the gear to be cut and n the number of teeth in the index worm wheel.

When multi-start hobs or index worms are used the expression becomes $\dfrac{DR}{DN} = \dfrac{N}{n} \times \dfrac{s}{S}$

where S and s = the number of starts in the hob and index worm respectively.

It becomes necessary in practice to provide other facilities on the machine which are not shown in Fig. 3, such as provision for adjustment of the relative hob and work positions, feed mechanisms, hob speeds

and suitable interlocks to prevent simultaneous engagement of con-
flicting motions.

The effect of these variables on the kinematic train is considerable,
but the increase in the number of gears and shafts can be dealt with by
introducing a machine constant. The index train then becomes

$$\frac{DR}{DN} = \frac{A}{B} \times \frac{C}{D}$$

$$= \frac{M_c \times S}{N}$$

A, B, C and D are the number of teeth in the index change gears and N
is the number of teeth in the gear to be cut. S is the number of starts in
the hob and M_c the machine index constant.

The direction of feed of the hob relative to the gear can be achieved
in any one of three ways but these basic methods can be combined to
form other permutations. The most usual forms are shown in Fig. 4.
The machine-feed constant is the distance which the hob slide will
advance during one complete revolution of the work spindle if the ratio
of the feed change gears is 1:1.

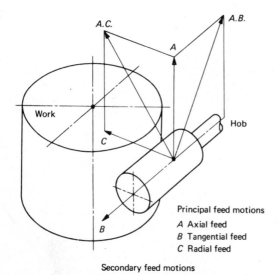

Principal feed motions

A Axial feed
B Tangential feed
C Radial feed

Secondary feed motions

A.B. Combined axial/tangential
A.C. Combined axial/radial

Fig. 4. Principal feed motions

B

To determine the feed change gears for hobbing spur gears simply divide the desired feed by the machine-feed constant and find suitable change gears for this ratio i.e.

$$\frac{DR}{DN} = \frac{\text{Feed}}{\text{Feed constant}}.$$

The feed is usually expressed in terms of inches or mm per revolution of the work table and is measured along the axis of the gear. When cutting helical gears, however, the hob is inclined to the helix angle and since the cut takes place along the tooth flanks the feed is given a multiplying effect by the helix angle. Thus if the feed gears are arranged to give feed f the actual feed along the helix will be $\dfrac{f}{\text{cosine helix angle}}$.

To avoid overloading the hob and to maintain a reasonable surface finish on the gear, the feed gears should be compensated to suit the helix angle. There is no specific relationship which must be held between the index and feed when cutting spur gears. When cutting helical gears, however, the index and feed must accurately be timed or a differential mechanism introduced into the gear kinematics.

Differential

The need for this device can best be seen if consideration is given to the case of finish cutting a helical gear which has already been rough cut. If a hob is centralized with a tooth space on the gear and the index gears arranged to give the necessary velocity ratio between hob and workpiece, when the machine is set in motion the hob tooth will remain central. However, if the vertical feed is then engaged so that the hob traverses through the face width of the gear then, owing to the helix angle, the hob tooth will start to contact one flank of the gear and depart from the other. This will continue if the vertical feed is kept in engagement until the tooth is cut completely away.

It follows that some form of compensation must be introduced to the feed and index motions such that the required helix angle can be produced. This compensation can be effected either by the differential or, on machines without differential, by modifying the index and feed gear trains. The differential is interposed between the index worm and the index change gears and provides the gain or loss of table rotation necessary to produce the helix angle.

Fig. 5 shows the relative ratio between a helical and spur gear. Both hob and gear are right hand helix and the face widths are equal to the

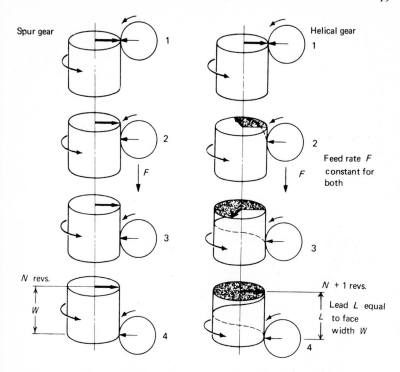

Fig. 5. Relative ratio between spur and helical gears

lead of the gear. In both cases the hob travels at the same rate of feed in inches per minute. The arrows at the end of the blank and the hob are reference lines which are established as shown in position 1. In the first position the hob is beginning to cut at full depth on both pieces and in position 2 the hob has fed one third of the face width of each gear, both hobs having made the same number of revolutions. The helical gear has made exactly the same number of revolutions as the spur gear plus one third of a revolution. Position 3 shows the hob two thirds of the way across the face and the helical gear has made two thirds of a revolution more than the spur gear. By the time the hob has reached the end of the cut the helical gear has made one more complete revolution than the spur gear.

Basic types

Hobbing machines are available in a number of forms, sizes and configurations, although the principle remains the same. Probably the most popular arrangement is the vertical machine where the axis of the gear to be produced and the feed motion are vertical. It is suitable for most sizes of machines from small to large diameter and makes for easy work handling. A typical example is Fig. 6, which shows the latest of the Churchill range, designed with special rigidity in the structure to allow the conventional tie-bar to be avoided and thus to give greater access for loading. The horizontal configuration is particularly useful for shaft work since by varying the length of the bed and hob traverse long shaft components can be handled with ease.

Fig. 6. PH1612 universal gear hobbing machine

Although most of the universal machines are capable of cutting fine pitch gears special machines are available which are designed for this type of work. Apart from the obvious economies obtained by restricting the diameter capacity of the machine, the kinematics are slightly different in that the fine pitch machines are capable of cutting large numbers of teeth. A 100 mm diameter capacity machine cutting a gear 0.14 mod. would have to be capable of cutting up to 700 teeth, and so the requirements for index ratio of the machine would be in the region of, say, 6–700:1. The larger universal machines are more likely to have an index ratio capacity in the region of 6–400:1, although this varies considerably with the manufacturer.

A further type of hobber is the high production machine which is usually limited in features in comparison with the universal model. It is different in kinematics in that hobbing is usually carried out at high speed with climb cutting. Hob speeds are usually much higher, up to say 500 rev/min, and index ratios may be more in the order of 6–200:1.

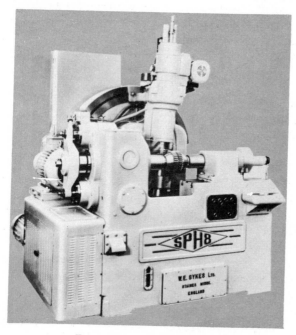

Fig. 7. High production hobber

Again, they vary considerably. Fig. 7 shows a high production machine once manufactured under licence by W. E. Sykes and still manufactured in the U.S.A. by the Michigan Tool Company. It is worthy of mention that the means of obtaining the helical action is unusual – no differential is required and the lead of the gear is obtained by means of a helical guide. The disadvantage of this method is that the guide must be changed for each number of teeth in the gear, i.e. for a given helix angle the lead changes with the diameter.

Under these conditions, i.e. 500 rev/min hob and 6:1 index ratio, the table would rotate at high speed, $83\frac{1}{3}$ rev/min. Such machines would therefore require carefully designed anti-friction table bearings and index drive. Rapid clamping, automatic loading and automatic swarf removal are typical features designed into such machines.

The turbine hobbing machines are usually the extreme high precision type, the gears produced running at high peripheral speeds necessitating considerable accuracy. Consequently this type of machine may be built as large as 5 metres diameter or bigger and modern machines have some form of built-in compensation for indexing inaccuracies. It is advisable to use one machine for large wheels and another for small pinions where production quantities warrant the expense. When using a large machine for cutting pinions it is necessary either to run the hob slowly, which does not give economical production, or to run the table worm at high speed, which is undesirable.

1.1.3 Hobbing techniques

A number of techniques are available for producing a variety of components and some of these techniques can be described as follows:

(a) Axial, tangential and radial feeds

These are the principal feed motions and are best illustrated in Fig. 8.

The most widely used is the axial method, where the hob is set to the required tooth depth clear of the blank and is then traversed in a plane parallel to the axis of the work. This necessitates an allowance for approach in a similar manner to a milling cutter.

The radial feed method is useful if the approach distance required by the axial method is limited since the hob feeds slowly into depth at a controlled rate per revolution of the work.

In producing wormwheels this motion is all that is necessary and the work is finished once the hob reaches its full depth. For producing gears which must operate at any position across the face width, however, the axial feed must now be engaged in order to generate the gear at all sections. Distributor gears on cam shafts are sometimes produced by plunge hobbing (radial feed) owing to the fact that the adjacent cams are too close to allow the hob to be traversed across the face width of the gear. This necessitates the mating gear's being set exactly on the centre line of the face width used when hobbing.

Tangential feed is also limited to certain applications but is the most accurate way of producing wormwheels, particularly since radial feed is not satisfactory for wormwheels of high helix angle. By moving the worm axially in relation to the wheel they can be made to operate as a rack and pinion since the tooth action is conjugate. Any section through the pair other than through the centre line of the worm will show that the rack section is asymmetrical. Sections taken at increasing distances

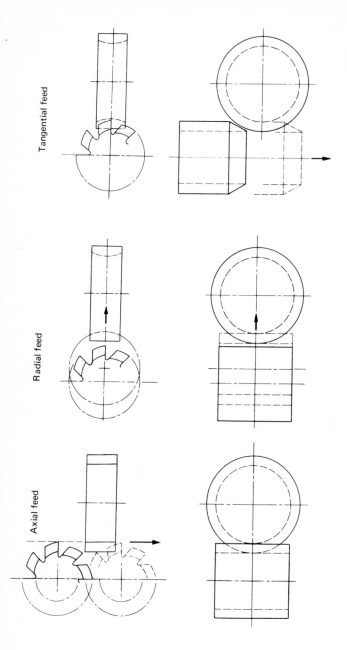

Fig. 8. Hobbing techniques

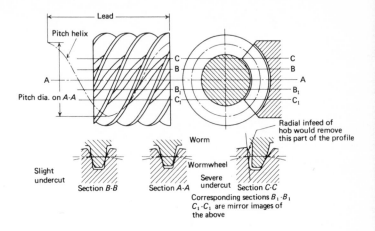

Fig. 9. Example of badly designed worm drive showing how undercut occurs on the off-centre sections when the lead angle exceeds the axial pressure angle.

from the centre line of the wormwheel, A, B, $C - A$, B_1, C_1 (Fig. 9) show that the pressure angle decreases until, at a point where it becomes zero, undercutting starts to occur on the leading flank on one side of the centre line and on the trailing flank on the other side of the centre line. When this happens radial infeed of the hob will remove the overhanging portions of the flanks with serious loss of contact area and the use of the tangential feed method becomes necessary. These undesirable features occur when the lead angle exceeds the axial pressure angle and mean that the worm must slide axially into engagement with its wormwheel.

The movement of the hob along its own axis changes the position of a given hob tooth relative to the work at each hob revolution, and as the movement is continuous it has the effect of multiplying the effective number of flutes. This apparent increase in the number of flutes makes the number of facets formed on the gear profile larger, thus materially improving the finish.

The differential causes the work table to rotate slightly faster or slower than is necessary to produce the required index ratio and to give the hob an equal additional axial motion. The tangential method of wormwheel production is slower than the radial method but the flute-multiplying effect is so great that it is possible to produce accurate wormwheels even when using the process known as 'flycutting'.

(b) Flycutting

When the cost of a hob is not warranted, or if time does not allow a hob to be manufactured, a fly tool can easily be made by mounting a high speed steel tool bit on a suitable body to give a hob with one tooth. The tooth should be inclined at the helix angle and suitably relieved behind the cutting edge to avoid rubbing.

If the worm is single start and the wheel has an even number of teeth, then for every revolution of the wheel the fly tool will engage in the same tooth space. By the time it re-engages with the same space, however, it will have advanced slightly owing to the tangential feed and the generating action on the tooth flank of the wheel is built up in this manner — the finer the feed the finer the generating action.

If the wheel does not have an even number of teeth then a 'hunting' action is obtained and the tool must be left to hunt out.

When the worm is multi-start the fly tool must be indexed or provided with teeth equal to the number of starts. For example, a two-start worm would require a fly tool with two teeth set accurately at 180° to each other, or if a single tooth tool were used the tool would have to be moved along its own axis a distance equal to half the lead once every other tooth on the wormwheel had been completed.

(c) Worm generating

Multi-start worms can be considered as spiral gears of low numbers of teeth and as such can be hobbed in the conventional manner. When the number of starts is less than four, however, it becomes necessary to set the index ratio of the hobbing machine (using a single start hob) to 4:1 or less, and very few machines are capable of operating at that ratio. Worms of one, two or three starts are best produced therefore by worm generating on the hobbing machine using a helical gear shaper cutter. The process is fast and accurate on the smaller worms but is limited on the larger pitches owing to the number of teeth in simultaneous contact while generating is taking place. Fig. 10 shows a typical example. The workpiece to be produced is placed on the hob arbor and the cutting tool on the arbor normally reserved for the workpiece. The machine is set as if the workpiece were producing the tool, i.e. the motions are reversed. The tangential feed head is required since the worm must be moved slowly along its own axis to complete the generation. There is no vertical movement of the hob head up or down the column and indeed care must be taken to see that the cutting edge of the cutter is set on the centre line of the worm. Better results are sometimes

Fig. 10. Worm generators

obtained with regard to cutting action and finish by setting the cutting edge just below centre but this produces a slight change of form on the worm and should therefore be done with discretion.

One of the difficulties or limitations of worm generating lies in the sharpening of the cutter since the cutting edges are usually involutes in the transverse plane. If the cutters are low in helix angle, say 15° maximum, they can be flat-face sharpened as if they had straight teeth. Higher angles, however, give considerable trouble since when sharpened in this manner the cutters have a negative rake on one flank which produces a poor finish on the component. If they are sharpened at right angles to the helix it is virtually impossible to maintain the involute profile on the cutting edges. Fortunately, the problem sometimes resolves itself since the worms with higher helix angles are usually more than four starts and can therefore be hobbed with a single start hob.

Other limitations of this process are in regard to the fillet produced in the root of the worm by the tip of the cutter. If the worm is generated by a 75 mm diameter cutter and is designed to run with a worm wheel of say 250 mm diameter, then care should be taken to check that the tips of the wormwheel teeth do not foul the fillets generated by the tip of the gear shaper cutter. The second limitation is the position of the base circle of the cutter relative to the addendum of the worm. This is

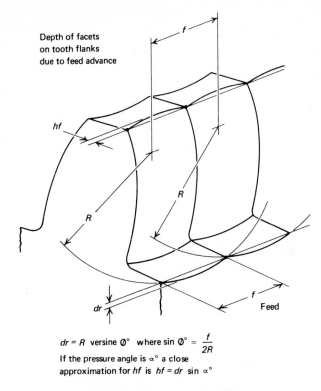

Depth of facets
on tooth flanks
due to feed advance

$dr = R$ versine $\emptyset°$ where sin $\emptyset° = \dfrac{f}{2R}$

If the pressure angle is $\alpha°$ a close
approximation for hf is $hf = dr$ sin $\alpha°$

Fig. 11. Feed facets — gear hobbing

shown in Fig. 171 and is discussed in detail in Section 2 Part 2.2.4, 'Worm generating cutters'.

Many modern high production hobbers are capable of operating at feed rates as high as 6 mm per revolution and this leaves a coarse feed pattern along the flanks. The visual effect is usually worse than the actual error and the depth of the feed scallops is in fact quite small and they are easily removable in the finishing operation. Fig. 11 shows the effect of the feed rate on the teeth along both the flanks and the root. For example a 100 mm diameter hob operating at 2½ mm feed would produce facets approximately 5 to 7 microns deep in the plane dr and 2½ microns deep in the plane hf. A 100 mm hob at 6 mm feed would produce 90 microns dr and 30 microns hf. Obviously these coarse feed rates are not intended for finish hobbing but for gears which are to be finished by some other medium.

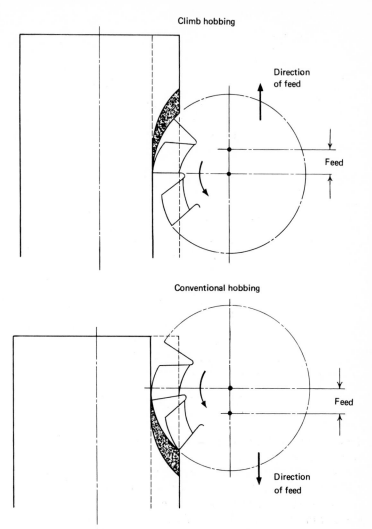

Fig. 12. Climb/conventional hobbing

(d) Climb/conventional hobbing

Fig. 12 shows the two conditions and the type of chip produced. The climb method usually gives a better finish on the work and also improved tool life, although there are sometimes conditions where the reverse applies.

It is generally agreed that since the hob teeth get well below the surface at the start of the cut there is less chance of rubbing occurring at the tips of the teeth and the greater pressure occurs at a point a little distance away from the extreme edge. Conventional hobbing usually takes place at conventional cutting speeds whereas climb hobbing is nearly always used with high speed hobbing, though again there are exceptions to the rule.

(e) Diagonal hobbing

This is a refinement of the tangential and axial feed methods, which means that the hob moves tangentially along its own axis in the direction

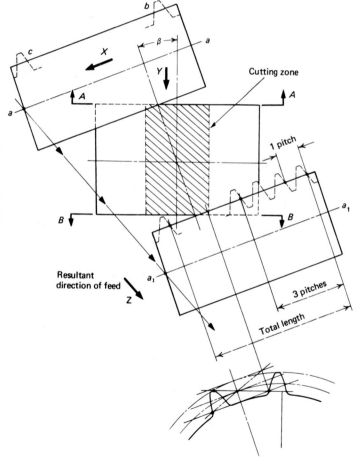

Fig. 13. Diagonal hobbing techniques

of arrow X (Fig. 13), while at the same time it moves axially in the direction Y across the face width of the gear. The resulting combined action is a diagonal movement in the direction Z, which has the effect of bringing virtually all the hob teeth into action when cutting a gear. The wear on the hob teeth is thus distributed over the length of the hob and a continuous hob shift is obtained. Fig. 14 therefore shows how the wear on the hob teeth can be distributed by imparting a motion along the axis of the hob while traversing through the face width of the gear. Fig. 14a shows a hob operating in the conventional manner down the axis of the gear i.e. axial feed, and it can be seen from Fig. 14d that all cutting takes place in the zone $ab\,a_1\,b_1$. The contact point x_1 between the hob flank and the gear does not change and it is necessary to introduce hob shift at the end of each cycle to ensure that the heavily loaded areas are not used again in the same place.

Fig. 14b shows the gear being produced by a further method (non-differential techniques, discussed on page 35) which also involves dis-

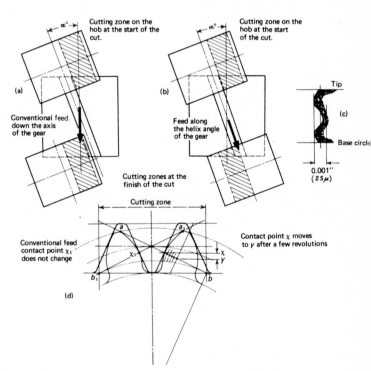

Fig. 14. Conventional and diagonal hobbing

placing the hob along its own axis relative to the gear axis. It can be seen that the contact point on the hob flank is now continuously changing — from point x to y in a few revolutions. The advantage of this type of technique is therefore obvious — the wear on the hob teeth is distributed and a continuous hob shift is obtained. The disadvantage of the diagonal technique, apart from the fact that a special hob head is required, is that all the errors in the hob are imparted to the gear in each cycle. Under normal conditions, i.e. axial feed, only some 1½ pitches of hob teeth would be necessary to generate the gear, therefore only the adjacent pitch error of the hob would be of any importance. With the diagonal technique, however, anything between three and six pitches could be used depending on the application, and it is thus necessary to consider the accumulative pitch error in the hob over say six pitches, which is obviously larger than the error in one pitch. Fig. 13 shows that at the start of the cycle, the hob axis aa is in such a position that the first teeth b on the hob are cutting the end section AA on the gear. At the end of the cycle the hob axis has moved to position $a_1\ a_1$, and now the hob teeth c are producing the end section BB on the gear. The involute on the section AA of the gear has been produced by the convolution b on the hob and the errors in this convolution only are imparted to the gear. The section BB on the gear has been produced by convolution c of the hob and the errors in this part of the hob only are imparted to the gear. Since the hob is accurately manufactured so that any single convolution in its length does not exceed a given tolerance, it follows that the involutes at AA and BB should correspond to each other within that tolerance. The errors in each single convolution of the hob can accumulate, however, such that the error between teeth b and c of the hob could be considerable. For example, if a 6 mod. hob is made to Grade A limits (B.S. 2062, Part 1), the allowable error over one convolution of teeth is 20 microns, while the tolerance over three convolutions is 30 microns. By taking advantage of the tolerance it would be possible to have a tolerance of 60 microns over six convolutions (although any single convolution would be within 20 microns).

It can be seen, therefore, that when the teeth c of the hob reach the plane BB on the gear they can be as much as 60 microns out of the required theoretical position for the example stated. In calculating the change gears for the motions in directions X and Y the theoretical values are used to determine direction Z, which can lead to considerable inaccuracies. The inaccuracies in the gear produced by this method do not go up in direct proportion to the errors in the hob, since this proportion depends on how the errors in the hob are distributed, but they will increase unless the hob is made so that the errors do not accumulate.

(f) Oblique hobbing

The previous method involves the use of a special hob head capable of feeding in two directions at once, and an extra motion is imparted to the differential to compensate for the extra motion obtained by feeding the hob along its own axis. A refinement of the diagonal method is the technique sometimes referred to as oblique hobbing, which has the same advantages but not the same disadvantages. Figs. 14 and 15 show that the hob head is mounted on a separate feed slide which can be inclined at the helix angle of the gear. The straight line generators of the hob therefore sweep across the face width of the gear at the pitch line helix angle and the motion is extremely accurate since no change gears or differential are required. The hob head can be set very accurately by means of a sine bar and small changes in helix angle can easily be effected.

It can be seen that if the ratio of the tangential and axial feed motions in the diagonal technique is chosen so that the resultant feed motion is down the pitch line helix there is no difference between the

Fig. 15. Hobbing machine equipped for diagonal feed

two methods. The oblique technique is best suited for the special high production machines used in the automobile industry where the gears are subsequently finished by a further process and the face widths of the gears are small. The helical overlap through the face width of the gear is usually approximately one. The number of hob thread convolutions brought into play to produce a gear is therefore of the order of two to two and a half. This number is well controlled on the hob and is not subject to excessive cumulative pitch error.

It should also be appreciated that the nature of the error produced on helical gears by the diagonal and oblique processes is quite different from that produced by the conventional axial feed method.

On the high production machines producing relatively small gears the principal errors produced are the result of hob rotational frequency. This type of error results in profile errors on the flanks of the gear teeth and its amplitude depends on the accuracy of the hob and its mounting and the concentricity of the driving elements. If errors are present the motion of the cutting surfaces during each revolution of the hob does not correspond to the angular motion of the gear and a complete phase of error takes place during each hob revolution. Figs. 14a and 14d shows that the contact point between a given hob tooth and the gear flank is at x_1 and does not change with the axial method. Hob rotational errors cause over-cutting and under-cutting of the gear flanks so that the resulting involute profile appears as shown in Fig. 14c. It also follows, therefore, that since the contact point x_1 of a given hob tooth does not change, the points of maximum profile error on the gear remain the same relative to the tip of the gear. When considering the diagonal techniques, however, it can be seen from Fig. 14b that the contact point between hob and gear is continuously changing and therefore the points of maximum profile error are also changing and move along the line of action. The result is shown in Fig. 16a, which illustrates the error characteristics obtained by axial feed. The profile at various sections a to e is indicated and it can be seen that the points of maximum profile error do not change their position relative to the pitch line or tip of the tooth. Since the lead error is measured along the pitch line XX it can be seen that the position of the profile error does not influence the lead accuracy. Fig. 16b shows the same gear produced by diagonal or oblique feeds and it can be seen that the points of maximum profile errors change at each axial section a to e of the face width. When the lead is checked along the pitch line XX, however, the profile error on sections b, c, d and e now has an effect on the lead recorded. Fig. 16c shows the

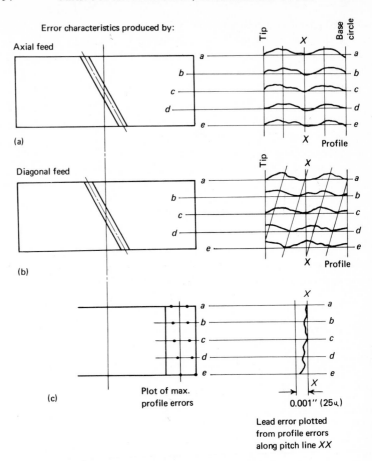

Fig. 16. Error characteristics produced by axial/diagonal feed

lead error plotted from the profile errors at the various sections along the line *XX*.

Since the oblique method is essentially a semi-finishing technique and the gears so produced are subject to a finishing operation, the apparent lead error is corrected as the profile is refined. It is questionable whether the gears produced by the oblique method are any better or worse than those produced by axial feed but the displacement of the errors along the flanks is certainly different.

(g) Non-differential hobbing

Machines which are not equipped with a differential mechanism can still be used to cut helical gears by suitable calculation of the feed and index gear trains.

The lead of the gear is the axial length in which the tooth spiral makes one revolution. If we consider the case of a spur gear having a face width equal to the lead of the helical gear to be produced then we have:

Spur gear face width $= F$

Helical gear lead L = Face width of spur gear $= F$.

Assuming the feed rate per revolution of the table to be f then the table would make $\dfrac{L}{f}$ revolutions whilst the hob traverses the face width F of the spur gear.

In the case of the helical gear lead L, face width F, the table will make the same number of revolutions: ± 1 according to whether the gear and hob are of the same or opposite hand of helix.

Thus the table makes $\dfrac{L/f}{L/f \pm 1}$ revolutions when traversing a face width equal to its lead.

Let $a = \dfrac{L}{f}$

Therefore: $\dfrac{a}{a \pm 1}$ = number of revolutions of the table = $\dfrac{1}{1 \pm \dfrac{1}{a}}$

As the helix angle decreases, the lead increases, reaching a value of infinity for a spur gear; then a becomes infinity and the formula becomes 1 indicating that the normal indexing ratio is required. The difficulty with this method is that the formula frequently results in a ratio that is difficult to obtain with available change gears.

It can be made more variable by changing the feed rate and indeed this then becomes necessary before a workable combination can be found.

Assume a machine index constant of K.

Therefore: $\dfrac{Kn}{N} \times \dfrac{1}{1 \pm \dfrac{1}{a}}$ = Index ratio

where: N = number of teeth in the gear
and n = number of starts in the hob.

The positive sign is used where the hob and gear are of opposite hand and the negative sign where they are of the same hand. The table indicates the sign to be used for any given set of conditions:

Hob	Gear	Type of cut	Sign
R.H.	R.H.	conventional	−
R.H.	R.H.	climb	+
R.H.	L.H.	conventional	+
R.H.	L.H.	climb	−
L.H.	L.H.	conventional	−
L.H.	L.H.	climb	+
L.H.	R.H.	conventional	+
L.H.	R.H.	climb	−

It follows that if the machine is equipped with a differential a great deal of tedious calculation can be saved, plus the fact that the feed rate can then be modified if required without having to recalculate new gear trains for the index.

(h) Prime numbers of teeth

The differential can also be used when cutting spur gears having prime numbers of teeth in excess of 100.

The index gear ratio is first calculated as close to the required ratio as possible and the helix so produced is corrected by a set of lead gears in the differential train cutting a helix equal to the difference between that produced and the required value, but of the opposite hand.

Thus N = required number of teeth and n = number of starts in the hob

$$\frac{Kn}{N_1} = \frac{DR}{DN} = \text{Index ratio}$$

Here N_1 is a number close to N which need not necessarily be an integer. It is then necessary to determine the fraction of a table revolution gained or lost in the number of hob revolutions required to produce the correct ratio.

Let this fraction gained or lost be A.

$$\text{Then } A = \frac{N_1}{N} - 1.$$

This applies when N_1 is greater than N and in this case the differential must add.

When N_1 is less than N, $A = 1-\dfrac{N_1}{N}$ and the differential must subtract. (This applies to right hand hobs and for left hand ones they must be reversed.)

The required differential ratio to correct this helix error can be determined from the formula $\dfrac{1}{A} \times f = L$

where L = lead
and f = feed.

The standard differential formula will depend on the machine constant for the differential, but assuming this to be K we then have

differential $= \dfrac{K}{L} \times \dfrac{N_1}{n}$.

By substituting for L the required differential gears can be found.

(i) Helical gears with prime numbers of teeth

When hobbing helical gears with prime numbers of teeth over 100 first calculate the main index gears to the nearest suitable ratio. Again, as for the last example, determine the fraction of table revolutions gained or lost and select the required feed gears.

Determine the differential ratio necessary to correct the indexing error and calculate the ratio to suit the helix angle.

The final ratio will be the algebraic sum of the ratio required to correct the index error and the normal differential ratio for helical gears.

(j) Cutting low helix angles

It may occasionally be necessary to cut gears having a very small helix angle and difficulty can be experienced in obtaining a gear train of the required degree of accuracy.

The following method will enable a wide range of helix angles to be produced within a fine degree of accuracy.

Calculate a set of gears for producing helical gears with a lead of rational proportions but of opposite hand to that required.

Find the differential train for this lead and calculate an index train that will by subtraction modify the initial lead to that required.

It is possible with this method to cut a gear with a lead as small as 25 microns per 25 mm of face width.

Gears of this description can be used in applications where the torque is high and continuous; the higher helix avoids the teeth bearing at one end only. This method is also suitable for cutting a bias into a gear prior to heat treatment in order to allow for subsequent unwinding of the helix.

Assuming a helix angle α_1° greater than the required helix angle α° and of the opposite hand, from this calculate the corresponding lead L_1.

Thus: $L_1 = \dfrac{\pi \times \text{P. dia. gear}}{\tan \alpha_1^\circ}$.

From L_1 find the gear ratio for the differential and let the differential ratio equal R_1.

Then $R_1 = \dfrac{K}{L_1} \times \dfrac{N}{n}$ (as determined by the formula on page 37).

If difficulty is experienced in obtaining this ratio with the available gears choose a ratio R_2 near to R_1 and calculate the resulting lead L_2 and helix angle α_2°.

Thus: $R_2 \qquad = \dfrac{K}{L_2} \times \dfrac{N}{n}$

and tangent $\alpha_2^\circ \quad = \dfrac{\pi \times \text{P. dia. gear}}{L_2}$

whilst $L_2 \qquad = \dfrac{K}{R_2} \times \dfrac{N}{n}$.

Now calculate a new differential train for a helix angle $\alpha_3^\circ = \alpha^\circ + \alpha_2^\circ$.

Thus $\dfrac{DR}{DN} = \text{Index ratio} = K \times \dfrac{n}{N} \times \dfrac{L_3/F}{L_3/F \pm 1} \quad \left(\begin{array}{l} - \text{ same hand} \\ + \text{ opposite hand} \end{array} \right)$

$L_3 \qquad = \dfrac{\pi \times \text{P. dia. gear}}{\tan \alpha_3^\circ}$

and is the same hand as required in the final component.

(k) High-speed hobbing

This is a technique which has been in use now for a number of years and about which so much has been said and yet so little written. Probably the best reason for this is that it defies a theoretical analysis and needs practical tests carried out under controlled conditions to get any factual evidence.

Cutting speeds depend on the material to be cut but, up to the advent of high-speed hobbing, it would be reasonable to say that a free machining carbon steel would have been cut at approximately 30 metres/min, this having been found from experience to give the optimum hob life. If the speed were increased to, say, 45 metres/min then the hob life suffered and the rate of wear became so rapid that it was not economical to run at that speed.

It was found, however, that if the speed were increased again this critical peak for hob life was passed and the rate of wear started to reduce. In fact at speeds in the region of 55 metres/min, and then again at 77 metres/min, the optimum results in terms of hob wear were achieved once more. It appears, therefore, that there are perhaps two or three critical areas in which it is economic to run the hob. Unfortunately, so many factors affect the critical cutting speed that these figures of 55 and 77 metres/min can be regarded only as a very rough guide.

Not all machines are suitable for high speed hobbing. This technique involves high rates of metal removal; consequently the machine must have high static and dynamic stiffness and the tooling must be rigid. Any lack of stiffness in the machine, or in the work fixture, is reflected immediately in the tool life and general performance. Climb hobbing is used with high-speed hobbing since the best results are obtained when combined together (although high speeds have been used with conventional cutting).

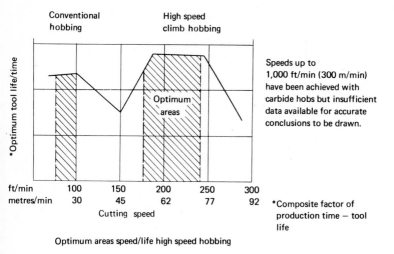

Fig. 17. Optimum cutting speeds

Fig. 17 shows the peak areas achieved with high-speed hobbing under a particular set of conditions. It cannot be regarded as true for the same conditions if used on a different machine, but at the same time it does indicate the critical areas and serves as a rough guide. Cutting force levels measured at Cambridge University are discussed in Section 4, and it can be seen from the traces of force levels measured in both climb and conventional hobbing that the forces are much smoother when the climb technique is used. This would appear to account for the tools' ability to withstand the increased speeds.

Perhaps another reason for this amazing increase in hob life at high speeds lies in the constantly changing cutting action of the hob and the size of the chip removed. It could be argued that at the lower speeds the hob tooth is in contact with the chip and the work for a longer period of time which results in more heat transferred to the hob thus causing wear. The high-speed technique means that the hob is not in contact with the chip for so long with the result that there is less heat transferred and the chips come off cooler.

Whatever the reason the undisputable fact is that the high-speed technique does give good results, although it must be realized that accuracy suffers owing to the high metal removal rates. The use of the high-speed technique can give production savings up to 100% but these have to be paid for and the result is usually a slight reduction in the accuracy produced. However as this method is used only when a further finishing medium is being employed this is no handicap.

1.1.4 Special applications and features

(a) Automatic two-cut cycle

Hobbing, unlike shaping, is basically a one cut process and different pitches and rates of metal removal are accommodated by varying the cutting speed or feed. Cases do arise, however, where it is necessary to take two cuts. It may be that (a) the gear to be produced is top capacity for the machine, (b) a high degree of accuracy is required, or (c) a high degree of surface finish is wanted, coupled with good productivity.

Two-cut hobbing meets these requirements, but to be economical each cut must be applied automatically rather than by the operator. The method of automatically obtaining the cuts varies with the design of the machine tool, but probably the easiest and most flexible way is where the radial infeed is controlled by means of a hydraulic cylinder. Such a device has a deadstop against which the hydraulic cylinder or piston abuts and this determines the finished depth of tooth, i.e. the last cut when considering multi-cutting. Other cuts are obtained therefore by controlling the movement of the hydraulic cylinder or piston and this is fairly easy to achieve. A machine with this degree of sophistication would also have facility of rapid traverse on the non-cutting motions.

(b) Variable feed and speed

Most modern hobbers can now be equipped with facility for variable feed and speed. This means that the feed and/or speed is stepless between a given maximum and minimum value and can be dialled in while the cut is in progress if necessary. The operator is thus given a chance to determine if the machine is cutting at its optimum rate or if the performance can be increased. It also gives the machine designer an opportunity further to extend the machine facilities by providing an auto change of the feed and speed rates. The least metal removal occurs

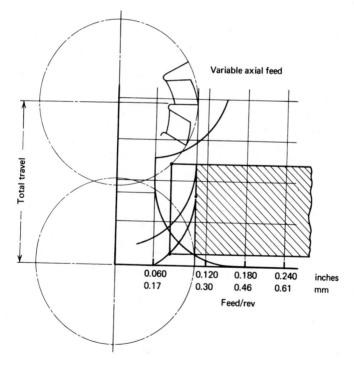

Fig. 18. Variable approach feeds

when the hob is in its approach cut; therefore the feed rate can be increased during the approach and gradually diminished to its required value once the hob is fully in the cut. This gives a further reduction in cycle time without adding undue wear to the hob, but it does require facility for varying the feed. Fig. 18 shows quite clearly the technique and the product savings can be considerable when the depth of tooth or approach length is large.

Another variant of the theme is the copying attachment. Although the object of this attachment is not to reduce the production cycle, it is closely allied to the above variants in that it relies on close control of the radial infeed to operate. It is far more sophisticated in that the close control must operate on two motions simultaneously and these motions lie in planes at right angles to each other. As the name implies the copying attachment enables the machine to copy or to reproduce a form from a master template. As the hob head feeds up or down through the face width of the gear the centre distance between the hob and work is modified by means of the radial infeed control.

The method of control varies with different manufacturers, but basically the stylus of the tracer head has its impulse magnified electronically and by means of suitable servo valves the radial infeed cylinder can be displaced as required.

Various programme cycles which can be pre-selected are now quite common practice on most up-to-date machines. One of the most start-ling at first sight is the ability to cut two different numbers of teeth at one setting. Once the limitations are stated, however, it can be seen that the problem is not quite so difficult as would first appear. The gears must be such that they can be produced with the same hob and must have the same helix angle and hand of helix. The difference between the diameters of the two gears must not be too large otherwise physical fouling points can be encountered on the machine. The change in diameter is accomplished quite easily as this sequence is similar to the automatic two-cut cycle and only involves two settings for the radial infeed device. The only other requirement now is to change the index gear ratio and this can be accomplished in several ways depending on the required ratios.

The easiest is to provide two sets of index gear trains with a suitable electro-magnetic clutch such that either train can be engaged as required.

Again these various cycles can be put together in combination such that when the first gear is complete the machine rapidly traverses to the second gear whilst changing the index gear ratio. This interrupted or stepped hobbing can also be utilized when cutting two or three sets of teeth of the same size since it does speed up the cycle and avoids the time that would be wasted if the normal feed were used when cutting air. These various cycles and programmes serve to indicate the high degree of sophistication available in the hobbing machines of today.

(c) Automatic hob shift

This device has now become more of an essential unit rather than a potential extra. Unfortunately, although many people realize its advantages, they do not understand the reason for the hob's wearing more at one point than at another and how to determine the optimum amount to shift the hob. When certain teeth of the hob are cutting a bigger chip is taken due to the position of the contact points between the hob and the gear flanks. As this contact point changes so does the size of the chip produced by a given tooth. While the hob makes one revolution the gear moves through one pitch, and (assuming there are F teeth in the hob) F cuts are made while the gear rotates through one pitch.

Most of the work is done by the tip of the hob and this usually occurs in some two, three or four consecutive teeth. The rest of the teeth of the hob cut on the side flanks only and the cuts are much lighter. These heavy cuts on the tips of the hob teeth cause the most wear, consequently some teeth on the hob show more signs of wear than the others. The object now is to move the hob along its own axis so that these heavily worn teeth are moved out of engagement and teeth not so heavily worn are used until they too are moved on. To find out the optimum amount to move the hob, trial cuts should be made to determine the number of teeth which are worn; in practice this is usually three.

Assuming twelve flutes in the hob, therefore, it is required to move the hob along its own axis a distance equal to

$$L \times \frac{a}{F}$$

where L = the lead of the hob,
$\quad\quad F$ = number of flutes in the hob,
and $\quad a$ = number of teeth badly worn.

(d) Chamfering starter rings

A very unusual application of the hobbing technique is the production of starter rings for cars and trucks. To allow easy engagement of the starter pinion the leading edge of the ring is chamfered, generally as shown in Fig. 118. This chamfer can be generated on a hobbing machine by means of special hobs (Fig. 159) which enable continuous indexing of the component and very fast time cycles to be achieved.

As the exact formation of the chamfer is not critical it is sufficient for one tooth of the hob to be used to produce it. If a one tooth hob were used (i.e. a fly cutter), the result would be acceptable since for every revolution of the hob the cutting tooth would produce the chamfer on one tooth of the gear. The process can be speeded up, however,

Pfauter combined hobbing/shaping technique

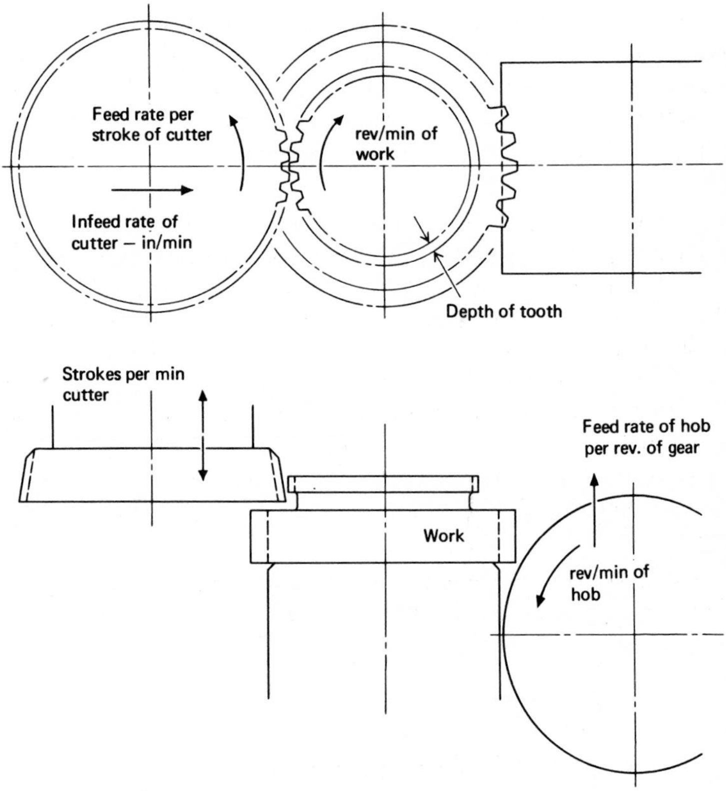

Fig. 19. Schematic diagram of combined hobbing and shaping unit

by using a number of fly cutters, that is to say, making the hob multi-start. If the hob were made with ten starts and ten flutes then there would be one start on each flute and, by clearing away all other teeth, the hob would have one tooth on each flute and one tooth for each start. It would in fact be ten fly cutters ganged together and therefore the gear would be cut faster as it would index ten times as fast (i.e. one revolution of the hob would equal ten teeth on the gear). Obviously the hob cannot be set immediately to depth — it must be fed in at a controlled rate until the desired amount of chamfer is produced. This can be achieved in one of two ways — either by radial infeed to depth with no axial travel of the hob, or by setting the radial depth and using the axial traverse.

(e) Combined hobbing and shaping

This technique was introduced by the Pfauter company of Germany a few years ago. The idea is, briefly, to combine a hobbing and a shaping unit into one machine so that certain types of component which would normally require two operations can be produced in one set-up. Typical of this type of component is the speed gear for a manual transmission as shown in Figs. 19 and 20. On this type of component the clutch teeth must be shaped and the helical gear would normally be hobbed and

Fig. 20. Combined hobbing and shaping unit

shaved. As conceived by Pfauter the shaping section is made as a complete unit which can be added to the basic hobbing machine, the drive for the shaper being taken from the end of the index wormshaft of the table.

This technique involves a considerable change from the conventional method of shaping gear teeth in that the gear is no longer produced in one, two or three cuts depending on the tooth depth. The hobbing machine is set in the usual manner and the speed of the hob and the index ratio determines the speed of rotation of the table. This would be considerably faster than would be achieved by conventional gear shaping under normal conditions and therefore it is necessary to revise the rate of feed per stroke of the cutter. This feed rate now increases tenfold approximately, depending on the conditions. Therefore it is obvious that if a reasonable size of chip is to be maintained then some other factor must be reduced. The only other factors which can be changed are the depth of cut and the infeed rate. Since the rotational speed of cutter and work are now so high the infeed rate drops to approximately one tenth of its normal value, and the number of cuts becomes very large such that infeed is continuous until the full depth is reached. In this manner the size and shape of the chip produced are changed drastically but the amount of work performed per minute is similar to that achieved by conventional gear shaping.

The cutter reciprocates along its own axis in the usual manner but at the same time rapidly rotates around the axis as it is fed slowly in to depth. Once the full depth has been reached the table makes several more revolutions in order to finish the teeth. For the example shown in the figure, the gear data are as follows.

Clutch	30T	10/20 D.P.	3 mm depth of tooth
Helical Gear	29T	23.7° H.A.	8 C.D.P.
Hob	75 mm diameter	400 rev/min	3 mm feed
	Cutting time 1½ minutes.		

Therefore the table speed

$$= \frac{1}{29} \times 400 \text{ rev/min}$$

$$= 13.8 \text{ rev/min}$$

In this time the table would make

$$\frac{90}{60} \times \frac{1}{29} \times 400 \text{ revolutions}$$

$$= 20.7 \text{ revolutions}$$

Cutter 100 mm pitch diameter   12 mm stroke
800 strokes per minute   40 teeth;

table makes, say, eighteen revolutions while feeding to depth; cutter does 800 strokes/minute for 1½ minutes (1200 total strokes).

Therefore the depth of cut (3 mm) is reached in 1200 strokes which gives an infeed rate of 2½ microns per stroke.

If the table makes eighteen revolutions to produce the gear then the

cutter makes $18 \times \dfrac{\text{teeth in gear}}{\text{teeth in cutter}}$ revolutions

$$= 18 \times \frac{30}{40}$$

$$= 13.5 \text{ revolutions.}$$

Since the total number of strokes to produce the gear is 1200 and the number of revolutions of the cutter is 13.5 it follows that it makes $\dfrac{1200}{13.5}$ strokes per revolution = 88.9.

The pitch circumference of the cutter is 100 mm $\times$ π and since it makes 88.9 strokes per revolution it follows that the feed per stroke is $\dfrac{100 \times \pi}{88.9} = 3.6$ mm.

For the example given, therefore, the infeed rate would be 2½ microns and the circumferential feed 3.6 mm per stroke, approximately ten times the value normally used for conventional shaping; but the number of cuts would be virtually equal to eighteen, which again would be approximately ten times as many cuts as would normally be required. It can be seen, therefore, that the same amount of work is being done in roughly the same amount of time but the manner of execution is different and so also is the chip formation. At the time of writing it is difficult to say whether this technique has much to offer over the conventional process, but experiments are being carried out and it should soon be possible to reach a definite conclusion.

1.1.5 Setting the machine

The accuracy built into the machine and the cutting tool will be lost if due care is not taken at the setting-up stage in the hobbing operation. When setting up the hob carefully clean the taper and locating diameter of the hob arbor and spindle and fit to the machine, checking with a dial indicator for true running. Clean the end faces and bore diameter of the hob and the spacing collars and fit the hob to the arbor, locking up after engaging the tailstock or outer support bracket. The nut should never be clamped without the tailstock bracket in place as this could cause the arbor to spring out of line. The hob bosses should be checked to run true within 5 microns and the high points of run-out at each end should be in line. Any eccentricity will produce an error in tooth form which will be aggravated when the high points of the run-out are opposite on the two ends of the hob. Such errors can be corrected by loosening the nut, rotating the spacers and reclamping; any slight error in parallelism of the faces can be utilized to correct the alignment.

The question now arises as to whether it is necessary to centralize the hob tooth or space on the line of centres corresponding to the shortest distance between hob and work. This is important when cutting low numbers of teeth, particularly straight-sided splines, but is not valid for large numbers of teeth unless appearance is a consideration.

When cutting low numbers of teeth the number of facets formed by the hob in generating a profile is smaller than when cutting large numbers of teeth. If a hob tooth or space is not centralized it is possible for the facets to be located nearer the tip of the tooth on one flank than on the other. This is more a question of appearance since it will not affect the running of the gears but the effect will be exaggerated if the hob is not running true. Any machine which uses automatic hob shift will move the hob tooth off centre once it has shifted, unless the amount of shift is made a function of the lead of the hob divided by the number of flutes.

For general work on gears, therefore, it is not essential to centre the hob, but it is definitely advisable when cutting straight sided splines of six teeth or less. The reason for this is described in detail in Section 2 Part 2.1.2 on spline hobs.

The hob axis itself must now be aligned relative to the gear axis and consideration must be given to the hob helix angle. When a spur gear is cut the hob head is set at the helix angle of the hob and for helical gears it is set at the helix angle of the gear plus or minus the helix angle of the hob (depending on the respective hands of the helices). It is desirable for the hands of helix of hob and gear to be the same in order not to show up the backlash in the kinematic drive (which would be the case if they were the opposite hand). If the hands of hob and gear are opposite, a gear can be produced but probably the feed rate will have slightly to be reduced and a slightly inferior finish may result. The setting of the hob helix angle is not too critical and very slight deviations show very little change in the shape of the tooth produced.

The correct setting of the hob head relative to the workpiece for various combinations of hand of helix is shown in Fig. 21.

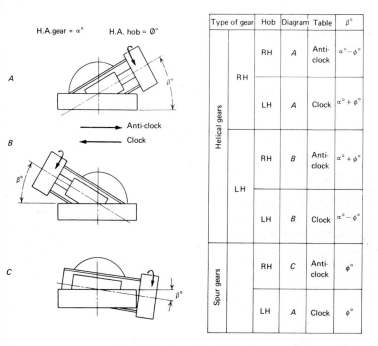

Type of gear		Hob	Diagram	Table	$\beta°$
Helical gears	RH	RH	A	Anti-clock	$\alpha° - \phi°$
		LH	A	Clock	$\alpha° + \phi°$
	LH	RH	B	Anti-clock	$\alpha° + \phi°$
		LH	B	Clock	$\alpha° - \phi°$
Spur gears		RH	C	Anti-clock	$\phi°$
		LH	A	Clock	$\phi°$

H.A.gear = $\alpha°$ H.A. hob = $\phi°$

→ Anti-clock
← Clock

Fig. 21. Setting of the hob head

The work-holding fixture plays an important part in determining the rate of production, degree of finish and accuracy obtainable. Too often it is assumed that given a good machine and accurate cutting tools good gears must result; unfortunately this is not true. Good gears can never be cut from bad blanks or inaccurate or poorly designed work-holding fixtures. Gear blanks should run true with the locating bore diameter and locating faces and should always be supported as close to the root of tooth as possible. This is particularly true of large diameter wheels but not quite so important on the smaller diameter pinions.

Time can be saved by multi-loading two or more blanks at one setting, providing the blanks are flat faced with no projecting bosses. The

D

approach and run-through time of the hob is now shared by a number of components and this reduces the cutting time per component. It cannot always be assumed, however, that this is necessarily desirable, for if a high order of accuracy is required it may be necessary to clock the blanks true and this would possibly take longer than the saving in production time.

In addition to this, the inaccuracies in the end faces of a component would slowly build up to a value which would be unacceptable.

The fixture used should be entirely dependent on the type of production involved (i.e., job, batch or mass production) and the degree of accuracy required.

Manually clamped, nut-type arbors are the cheapest but possibly the slowest type of work-holding equipment. They can be improved, however, by design as quick-change arbors, primarily suitable for small diameter parts and fine pitches.

For high production some form of power clamping, pneumatic or hydraulic, is almost essential. This can be cheapened by providing a power source as a permanent feature under the machine table and designing a series of mechanical parts which adapt into it by means of suitable draw bars. This kind of tooling is usually of the fail safe type, where the actual clamping is done by spring pressure and the power source is used to release the fixture. In this way if the power source fails the work remains clamped and thus does not scrap the component being cut. This power clamping can be applied in three ways:

(a)   friction clamping the end faces where reasonable areas are available;

(b)   the same, where limited clamping area is available;

(c)   expansion in the bore diameter of the component.

It should be noted that with the second type serrated driving faces are employed which lightly penetrate the face of the component thus giving the necessary grip to drive without slipping. This type of drive is not recommended for coarse pitch cutting.

For long shaft type work it may be necessary to employ some form of steady rest to stop resonance and vibration during cutting. This steady will register on some diameter on the component which is accurately controlled. Otherwise it is necessary for the operator to re-set the work steady for each component.

On the high production type of fixture it is usual to sequence it into the machine cycle such that after the operator had manually loaded the fixture he would press one button, or possibly two (depending upon the safety requirements), whereby the machine would automatically start its sequence once the fixture had clamped.

1.1.6 Calculating gear cutting time

Calculating the production time for gear hobbing is a simple matter once the basic mechanics of the operation are understood. The feed rate is usually given in terms of mm per revolution of the gear blank, and it is necessary to know the total number of revolutions made by the gear and the rev/min. Assuming:

rev/min of the hob = a

therefore rev/min of the table or work

$$= \frac{\text{number of starts in the hob}}{\text{number of teeth in the gear}} \times a$$

total distance travelled by the hob = b

where b = face width of the gear + overtravel and approach; if the feed rate is C mm per revolution of the gear then the gear makes $\dfrac{b}{C}$ revolutions during the full travel of the hob.

Therefore the time taken to cut the gear

$$= \frac{\text{total number of revs of the gear}}{\text{rev/min of the gear}}$$

Time taken

$$= \frac{b}{C} \times \frac{\text{number of teeth in the gear}}{a \times \text{number of starts in the hob}}$$

$$= \frac{b}{\text{feed rate}} \times \frac{\text{number of teeth in the gear}}{\text{number of starts in the hob}} \times \frac{1}{\text{rev/min}}$$

The overtravel varies with the circumstances (but it is usual to allow 3 mm), whereas the approach distance obviously varies with the depth of cut and hob diameter, as shown in Figs. 22a and b.

The above represents the solution for spur gears. Obviously with helical gears the situation is more complex since the hob is approaching the work at an angle. (The fact that the hob is inclined at its own helix angle relative to the spur gear has been ignored above since the exact calculation of the approach distance is not necessary.)

It is accurate enough for all practical purposes that the approach distance for a helical gear be calculated as shown in Fig. 22. It only remains, therefore, to determine the speed at which the material of the gear can be cut and also the feed rate to be used.

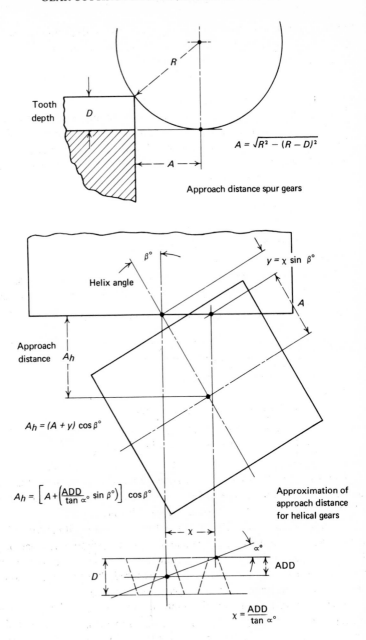

$$A = \sqrt{R^2 - (R - D)^2}$$

Approach distance spur gears

$$y = \chi \sin \beta°$$

Helix angle

Approach distance A_h

$$A_h = (A + y) \cos \beta°$$

$$A_h = \left[A + \left(\frac{ADD}{\tan \alpha°} \sin \beta° \right) \right] \cos \beta°$$

Approximation of approach distance for helical gears

$$\chi = \frac{ADD}{\tan \alpha°}$$

Fig. 22. Approach distance — gear hobbing

It is difficult to generalize with cutting feeds and speeds for different materials since much still depends on whether the gear is to be finish cut or finished by some other process, the condition of the machine and the type of tooling available. Assuming, however, that the machine is in good condition and that good tooling is available, we can restrict the number of variants. The following chart shows typical cutting speeds for various materials for finish cut gears:

Material	Cutting rate (metres/min)
Cast iron	25
Ph. bronze	36
EN8	28/30
EN352	25/28
EN24T	18/22

If the gears are to be high speed hobbed speeds in the region of 55 to 85 metres/min can be used; feed rate, however, is subject to more variables. The size of the machine and pitch of the gear dictate the rate which can be used. Obviously a 6 mod. gear can be cut with a higher feed rate on a machine capable of cutting 13 mod. than on a machine capable of cutting 6 mod. maximum.

A modern hobbing machine of, say, 4 mod. 175 mm diameter capacity intended for cutting gears for the car industry should be capable of operating at feed rates between 1½ mm and 3 mm per revolution of the work.

There are also instances of machines operating at feeds in the order of 6 mm per revolution but these are still rare and cannot yet be considered as typical.

Since the feed rate is always measured down the axis of the gear it is necessary to distinguish between the feeds for helical and spur gears. Helical gears are usually cut with a feed rate slightly less than would be used for a corresponding spur gear in order to keep the feed facets along the flank approximately the same.

If the tangential feed method is used a slight difference in the calculation for the number of revolutions of the table during the complete travel of the hob is necessary (see Fig. 23).

b = total distance moved by the hob

$e_1 = R \sin \beta^\circ$

d = depth of tooth

$$e = \frac{d - (R \text{ versine } \beta^\circ)}{\tan \beta^\circ}$$

α° = transverse pressure angle

$$\theta^\circ = \text{versine } \frac{d}{R}$$

Then total distance $b = R \sin \theta^\circ + R \sin \beta^\circ + \left[\dfrac{d - (R \text{ versine } \beta^\circ)}{\tan \beta^\circ} \right]$

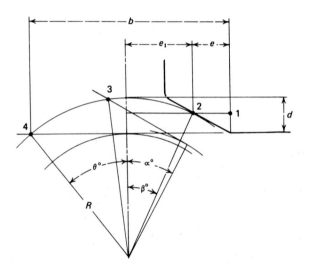

Fig. 23. Approach distance – wormwheels

If the component had no face width but consisted of a very thin lamination then the distance to be moved by the hob would be from point 1 to point 3 where generation would be complete. Owing to the fact that the component does have a face width which envelops the hob it is advisable to consider the hob as not having finished generating until it reaches point 4. Because of the enveloping action of the throat of the worm wheel the hob does not cease cutting until it reaches a position somewhere between points 3 and 4, the exact determination of which is difficult and complex and outside the calculation under consideration. As the above formula is only an approximation it may be considered more expedient to draw a single line diagram in order to determine the distance b.

Typical feed rates with this process would be in the region of 130 microns to 780 microns depending on the application. The number of revolutions of the gear during generation is therefore $\dfrac{b}{\text{feed}}$ where the feed is the tangential value.

If the radial infeed method is used, once again the calculation for distance b moved by the hob is slightly different. The hob now travels a distance equal to the tooth depth plus a reasonable approach distance of say, ½ mm.

The distance b therefore = depth of tooth + ½ mm.

Typical radial infeed rates are 130 microns to 400 microns per revolution of the gear.

Therefore the number of revolutions of the table to generate the gear
$$= \frac{\text{depth of tooth} + \text{½ mm}}{\text{feed}}$$

If a number of calculations for cutting times are to be made it is sometimes advisable to make up a small master record. This should include the feed rate in both normal and transverse planes for helical gears so that it can be clearly shown that some correction has taken place.

This record can be included in the gear-cutting operation sheet if required, but for clarity is best left as a separate document. The gear-cutting operation sheet should be as short and explicit as possible, so that the operator has no need to wade through a wealth of data which is of no interest to him in order to find data essential to cutting the gear. The sheet should include

(a) brief gear data to enable the gear to be identified at any time;

(b) measuring data to enable the gear to be checked and also machine settings for inspection equipment;

(c) outline sketch of the gear showing locating surfaces;

(d) fixture data and datum clocking registers;

(e) cutting tool data and gauges, if required;

(f) type of machine-feed, index, speed, differential gears, length of travel of hob, size of hob arbor;

(g) hob shift, if necessary — amount and frequency of shift;

(h) special notes (e.g. centralize the hob, type of coolant);

(i) degree of accuracy required;

(j) internal references — job number, etc.

1.1.7 The multi-start hob

So much has been written on this subject that it is difficult to be original or to provide fresh facts to conclude the argument one way or the other. Perhaps the most constructive approach is to collate all the facts in a logical manner without trying to provide a case either for or against.

Fig. 24 shows a one and two-start application. The work advances by one tooth for each hob revolution with the single start hob and by two teeth for a two-start hob. If a hob is 75 mm diameter and has twelve

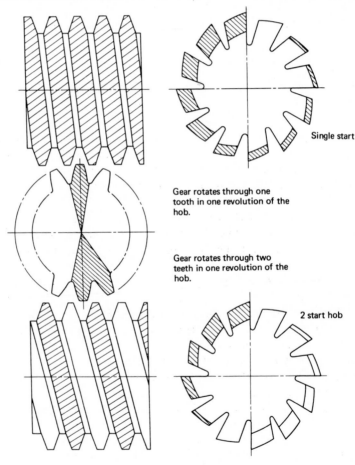

Single start

Gear rotates through one tooth in one revolution of the hob.

Gear rotates through two teeth in one revolution of the hob.

2 start hob

Fig. 24. Multi-start hobs

gashes for both the one and two-start applications, then there are twelve and six cutting edges respectively to form each tooth space. The chip load per tooth of the hob is doubled, therefore, for the two-start hob if the feed rate is kept to the same value as for the single start. Since there is no reason to suppose that one hob is capable of taking any more chip load than the other, it follows that the feed rate for the two-start application should be half that of the single start. If both hobs produce the same gear in the same time then they have removed the same volume of metal and it could be argued that they have had the same chip load per tooth.

We know from experience, however, that the multi-start hob produces the gear faster than the single start for the same tool life, although not in direct proportion to the number of starts. Although the chip load per tooth of the hob is the same for the same unit time on the gear, the shapes of the chips produced by the two hobs are different.

The single-start hob produces finer facets down the involute profile since it has more cutting edges available per tooth of the gear. However it also produces a coarser feed mark along the tooth parallel to the gear axis. If the multi-start hob were made with more cutting edges it would produce finer facets on the involute profile. Again, if it were possible to increase the cutting edges on the multi-start hob it would be possible to increase the single start which would then restore the *status quo*.

If the hob were increased in diameter it could be provided with more cutting edges, but an increase in diameter involves a reduction in hob spindle speed to maintain the same peripheral speed. A reduction in spindle speed would increase the cutting time for the gear and therefore no production saving could be achieved without an increase in feed rate.

If a large number of starts are used in the hob then the helix angle rises and in order to keep this to a reasonable value the hob diameter must be increased. This increase in hob diameter also brings in a further handicap in that the approach time to full depth becomes greater and brings a further rise in production time.

It can be seen therefore that

(a) Multi start hobs decrease the cutting edges available per tooth of the gear.

(b) If the cutting edges per gear tooth are maintained the hob must be increased in diameter and number of gashes.

(c) This increases the production time owing to the reduction in hob speed (rev/min) and increase in approach distance.

(d) Multi start hobs are more difficult to manufacture and as such are usually less accurate than single start.

(e) The work turns faster relative to the hob in direct ratio to the number of threads, which puts additional demands on the hobbing machine in that the index worm gear must rotate faster; this in itself could involve multi start index worms (depending on the application), and a slight deterioration in accuracy.

From this it would appear that there is no case whatsoever for the use of multi-start hobs. Certainly from the theoretical point of view more arguments can be prepared against than in favour. In practice, however, there is very little doubt that provided that they are used intelligently savings can be achieved. As the same amount of work has to be performed the only answer can be that the shape of the chip produced in multi-start hobbing allows high rates of production to be achieved. Fig. 25 shows the difference in the chip shape produced on the same gear by:

(a) a single start twelve flute hob and (b) a two start twelve flute hob.

The single-start hob is shown with 6.35 mm feed and the two-start with 3.17 mm feed. On the former it can be seen that the thickness of the chip t is nearly half that for the two-start but that the depth d is doubled. The only inference that can be drawn from this is that the smaller chip must be more conducive to better hob life.

Another factor which stands in favour of multi-start hobbing is the fact that owing to the faster indexing action the hob tooth is not in contact with the chip for so long and therefore less heat is generated. Again, if reference is made to Fig. 25 it can be seen that the heavy stock removal load falls on a few teeth in one area of the hob and the remaining teeth serve to finish out the flanks of the gear. If the hob is single start these few heavily loaded teeth fall on one side of the hob only which imposes a strong intermittent load on the kinematic drive. This produces the characteristic 'honk' of the heavily loaded machine.

If multi-start hobs are used, however, this load is split up around the hob periphery into two or more equally spaced locations depending on the number of starts. This tends to shade out the intermittent load effect. Thus if sufficient starts could be added the overlapping effect could produce a load which, although not constant, is smoother and better distributed and undoubtedly gives improved tool life.

So far we have considered the general situation but if we accept the fact that multi-thread hobbing has some merit, then we must consider

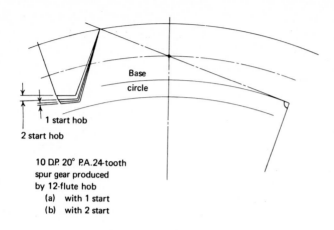

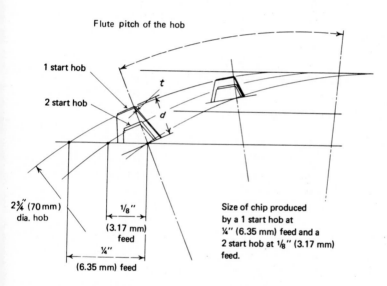

Fig. 25. Single/multi-start hobbing

the associated mechanical relationships between number of starts, hob gashes and the gear teeth. Individually these are not difficult to understand, but when combined together a permutation of variables can be obtained which requires careful analysis.

No matter how accurately the hob is made there is bound to be a difference between each individual thread and, depending on the relationship between starts in the hob and teeth in the gear, all the teeth may be formed by all the threads or certain teeth by certain threads.

There are two basic relationships to be considered — the number of hob starts to the number of gear teeth, and number of hob starts to number of hob gashes.

1 Relationship of number of hob threads to number of gear teeth

There are three possible relationships or ratios between the numbers of hob threads and the number of gear teeth — the ratio will be even. will have a common multiple, or will be prime. Different workpiece characteristics result from each of the three types.

The even ratio is the condition where the number of hob threads is evenly divisible into the number of gear teeth, as would be the case with a two-thread hob cutting twenty-two tooth gear.

Under this condition, a given hob thread will run in the same gear tooth spaces on every revolution of the work. Lead error and profile characteristics of certain hob threads will therefore be imparted to certain teeth as hobbing proceeds. Now, since just one thread has been associated with each gear tooth space, and as the thread on a hob is quite accurate, the profile produced on each tooth in even-ratio hobbing is also apt to be quite accurate.

The disadvantage comes in tooth spacing. Because of the inevitable minute lead errors among threads on the hob, the locating of these highly accurate teeth around the gear is likely to vary, with accompanying loss of spacing accuracy in the gear. It is generally found that the spacing errors of multi-thread hobbing are of greater magnitude and are more troublesome — harder to remove by finishing — than the profile errors. Also, as the threads are located at different points around the hob, any runout error in hob mounting will displace their gear-contact areas radially from one another. The result will be alternating high and low teeth around the gear which entails alternating variation in tooth thickness. This, in turn, will produce corresponding variation in backlash of meshing gears.

The common multiple ratio is the condition where, although the number of hob threads is not evenly divisible into the number of gear teeth, a number exists by which both are evenly divisible. An example would be a four-thread hob cutting a twenty-two tooth gear. The number of threads and the number of teeth share the common factor, two.

In common multiple hobbing certain groups of hob threads (the threads in the group being equal in number to the common multiple) run in certain groups of tooth spaces (the teeth in the group being equal in number to the common multiple). Thus there are sets of gear teeth, all of which have been affected by the combined action of two or more hob threads. Under such conditions, the minute profile imperfections of the different threads acting within the set will compound the profile errors on the teeth, but spacing discrepancies within the set will be largely eliminated because of the identical total thread exposure.

However, because each group of threads will differ slightly in profile and lead error characteristics from all other groups of threads, the alternate sets of teeth will vary. The profile errors are often able to be corrected by finishing operations. However, the inter-set spacing errors are of the widely spaced type that are difficult to remedy after hobbing.

The errors produced by runout in mounting are of the same general type as those found in the even ratio except that, instead of high and low gear teeth, you will have high and low sets of gear teeth. The operating problems are proportionately greater.

The prime ratio is the condition where the number of hob threads is not evenly divisible into the number of gear teeth, and there is no common factor. Such a case would be a four-thread hob cutting a twenty-one tooth gear.

In prime ratio hobbing the workpiece gains or loses, relative to the hob, with every revolution (there is always a 'hunting' tooth), so that by the time hobbing is completed, each gear tooth will have meshed repeatedly with every thread. The profile and lead error characteristics of all the threads are imparted to all the teeth, resulting in profile and spacing uniformity of the gear. This uniformity will be obtained in spite of any hob sharpening errors or mounting runout errors that may be present.

It should be noted that in the case of the generated profile, uniformity and accuracy are two different things. With, say, three or four different threads acting upon a gear tooth, profile accuracy will certainly not be improved. However, on the positive side, the prime ratio does afford the highest degree of tooth spacing accuracy available in multi-thread hobbing, and the profile errors are usually removable by shaving.

Also, though uniformity is obtained when hob mounting runout is present, other problems will be introduced by this condition. As in the other ratios, hob runout will still tend to produce high and low teeth. The circulating effect will cancel the difference between them but, in the process, the tooth thickness is reduced. The thread that cuts shallower leaves stock that is removed by the one that cuts deeper but the composite generating flat is quite large.

II Relationship of number of hob threads to number of hob gashes

A relationship that exists on the hob alone, this is the second major consideration in multi-thread hobbing. It determines the location of the generating flats and their sizes (although the sizing effect emerges only when the ratios of II are inter-related with the ratios of I).

As with case I, II consists of the three ratio possibilities, even, common-multiple and prime. As in case I, too, different workpiece characteristics result from each of the three, or at least from each of the three being mixed with one of the ratios from case I.

In the even ratio, the number of hob threads is evenly divisible into the number of hob gashes, as on a three-thread hob with eighteen gashes.

Under this condition, groups of teeth with identical generating flat structures are produced. It is possible to centre one hob tooth of each hob thread in a gear tooth space, which will result in gear tooth symmetry from the viewpoint of the generating flats. This can be an important consideration in pre-shave hobbing with a protuberance hob, where the exact location of the bottom-most generating flat on each side of the gear tooth may be critical for adequate shaving cutter clearance. As the number of gear teeth increases and as the pitch becomes finer, however, this problem decreases in importance.

In the common-multiple ratio the number of hob threads, though not evenly divisible into the number of hob gashes, shares a number by which both are evenly divisible. An example would be a four-thread hob with fourteen gashes, where the common factor is two.

Under these conditions, generating flats are alike on alternate gear teeth within certain groups, the number of teeth in each group being equal to the common factor. Only one hob tooth in one of the threads can be centred, meaning that the other relationships present will result in gear teeth that are asymmetric from the point of view of generating flats.

The prime ratio is the condition where the number of hob threads is not evenly divisible into the number of hob gashes and there is no common factor. This can be illustrated by the case of a hob with two threads and twenty-three gashes.

In prime ratio hobbing the generating flats are positioned differently on all the teeth and, as with the common-multiple ratio, only one hob tooth on one of the threads can be centred.

PART 1.2: SHAPING

1.2.1 Principle of the shaping process

Gear shaping is a generating process and the shaper cutter is virtually a gear provided with cutting edges. The tool is rotated at the required velocity ratio $\dfrac{\text{number of teeth in gear}}{\text{number of teeth in cutter}}$ relative to the gear, and any one tooth space is formed by one complete cutter tooth. Consequently any errors on a cutter tooth are reproduced directly in the corresponding gear tooth space. As the cutter rotates with the gear it forms the tooth space by various incremental cuts, depending on the feed rate used and the number of cuts taken to reduce the gear to size. Fig. 26 shows

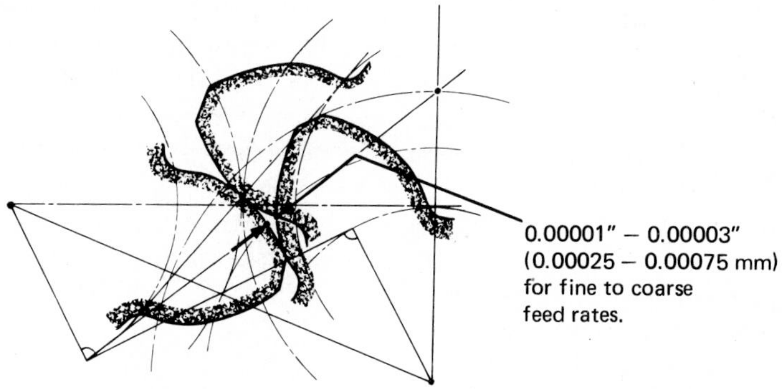

0.00001″ − 0.00003″
(0.00025 − 0.00075 mm)
for fine to coarse
feed rates.

Fig. 26. Feed facets − gear shaping

quite clearly that the teeth are formed by a series of closely spaced individual cuts and the involute on the gear is in fact a series of finely spaced cusps. The depth of these cusps is however exceedingly small, even for relatively high feed rates, and for all practical purposes the involute can be regarded as a smooth curve. Any indentations or tears on the flanks of the teeth are due to inaccuracies in either machine, tool, fixture or component and not to the feed rate or generating process.

Advantages and limitations

There are certain types of component which can be produced only on a gear shaper and others which, although they can be produced on a hobber, are best produced by the shaping method.

Racks cannot be processed by generating hobbing (they can be produced by an attachment using milling cutters but this is a form milling and not a generating process) but are rapidly and accurately produced by shaping. The maximum length of rack is only limited by the practical limitations of length of the table of the machine. Gears with adjacent shoulders which limit the run-out of the cutting tool are obvious applications for the gear shaping process. Although special attachments and hobs can be produced to allow internal gears to be hobbed, they are extremely limited in application and are nowhere near as versatile as the gear shaper.

The Sykes type double helical gear of the continuous or staggered tooth type can be produced only by the gear shaping process. Special components such as cams or irregular shapes and forms can usually be produced only by the shaping process. Worms and wormwheels cannot be made by gear shaping and apart from these limitations there are certain inherent problems which the machine designer has to face.

Owing to the reciprocating action of the cutter there is no cutting on the return stroke on single spindle machines. (This is overcome on the Sykes type double spindle machines discussed later.) Consequently only half the machine time is spent on removing metal. This reciprocating action creates severe problems for the machine designer owing to the reversal of masses and the large inertia forces which build up at high speed.

1.2.2 Principle of the shaping machine

Most single spindle gear shapers are of a vertical configuration though this is not an essential part of the principle – they work equally well in the horizontal plane.

Essentially the pinion type cutter is a gear with cutting edges and means are provided to rotate the cutter tight in mesh with the gear to be produced. To provide a cutting action the cutter is reciprocated along its own axis at speed and at the same time a feed advance motion is applied to one or other of the two members thus presenting different parts of the flanks of the cutter to the gear to form the involute profile. Fig. 27 illustrates the machine in its simplest form and as it is useful for the cutter and table index ratios to be the same the index constant is usually made equal to 1. Thus

$$\text{index} = \frac{\text{number of teeth in the gear}}{\text{number of teeth in the cutter}}.$$

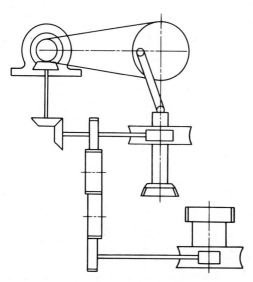

Fig. 27. Basic gear train of a shaping machine

The feed motion is measured as the advance per stroke around the pitch circumference of the cutter. Thus, in one revolution of cutter, it makes a number of strokes given by the ratio

$$\frac{\text{pitch diameter cutter}}{\text{feed per stroke}} \times \pi.$$

This value is often referred to as the F number and is used in the calculation for cutting times. It follows, therefore, that if the gear were the same diameter as the cutter, say 100 mm, the feed were 0.314 mm per stroke and the number of teeth in the gear and cutter were forty, then

$$F = \frac{100 \times \pi}{0.314}$$
$$= 1000.$$

Thus the gear has been produced in 1000 strokes of the cutter, or each tooth of the gear has been produced by twenty-five strokes of the cutter.

It can be seen, therefore, that an extremely fine finish can be built up by the shaping process, as the feed has a direct bearing on the number of cutting edges used to produce the involute, which is not so with the hobbing process. If the feed ratio in the previous example were

E

modifed to 0.0314 mm per stroke, any tooth of the gear would be produced by 250 strokes of the cutter. The pitch circle diameter of the cutter therefore affects both the index and feed ratios since altering the diameter changes the number of teeth and the F number.

The cutter is reciprocated along its own axis usually by some form of crank or lever arrangement, and the change of speed is accomplished by pulleys or gearbox. The path followed by the cutter is dictated by a suitable guide arrangement attached to one end of the cutter spindle. This guide unit can be changed according to whether spur or helical gears are to be produced. In this respect shaping is not so versatile as hobbing in that the guide must be changed for each helix angle or hand of helix. Other refinements must also be added, of course, particularly some means of feeding the cutter into the work, since unlike the hobbing process the cutter must be fed slowly into depth. The cutter is fed slowly into depth as the gear rotates. Thus the table has rotated through an angle before the depth of cut has been reached, at which point the infeed ceases. The means of applying the cut is therefore important and must be versatile since gears are produced in from one to four cuts, depending on the size of the gear and the machine tool itself.

Adjustment is also provided for the length of stroke and for the height of the cutter above the table or work fixture. One of the most difficult and important functions to be provided by the machine designer is that of relief to the tool during the return stroke of the cutter. The tool cannot be dragged back through the work in the cutting position since it obviously destroys the cutting edge and adversely affects tool life. Some manufacturers provide relief by moving the tool and others by moving the gear. The net result is the same but opinions differ slightly as to which is the better method. It is difficult to choose, as the effective relief depends on the manner of execution more than anything else. However one observation worth recording is that relieving the cutting tool may be an advantage in that at least the weight of the mechanism is constant so that the mass and inertia forces are known to the designer beforehand. Relieving the work means anticipating the largest weight ever to be placed on the table and allowing for the variation in the likely minimum and maximum values.

(a) Basic types

Shaping machines are available in a number of forms and configurations; probably the most popular is the vertical single spindle machine. The axes of the work and the cutter are vertical and this arrangement is suitable for most sizes of machine from small to large diameter and makes for easy work handling.

The precision fine pitch machines are available as vertical or horizontal, neither having any great advantage over the other. Possibly the vertical version is to be preferred since it may allow more accurate centring of the work. The horizontal type of machine is particularly useful for long shaft work or awkwardly shaped components since these can usually be loaded through the main spindle and auxiliary steadies for extra long work can conveniently be used.

In 1911 W. E. Sykes filed the first patents for a gear generator which finally overcame the difficulty of producing the continuous tooth double helical gear. A graduate of Leeds University, he went to the U.S.A. in the early 1920s and in association with Farrell Birmingham developed some of the largest gear generating machines of that time. He formed the present company that bears his name in 1927 and developed specialized manufacturing plant which pioneered the production of precision gear shaper cutters in Europe. In 1934 he achieved the world recognition he deserved when the Franklyn Institute of Pennsylvania awarded him the Edward Longstreth medal for his work on the design and development of machines for producing continuous double helical gears.

(b) Sykes horizontal machine

One of the main advantages of the horizontal machines is the ability to cut the double helical type of gear both of the gap and continuous tooth varieties. At first sight it would appear that the absence of tool clearance was an insurmountable difficulty when cutting continuous teeth but a combination of factors makes it possible.

The cutters are mounted with their cutting edges facing each other, and both cutter heads are arranged on one reciprocating carriage so that one cuts when the movement is in one direction and the other cuts when the direction is reversed. The two cutter heads can accurately be adjusted along the carriage so that each cutter will end its stroke when the cutting edges are at the apex of the teeth. If the tools had straight teeth and did not rotate then the reciprocating motion would be such that at the end of its stroke each tool would leave a small burr or chip which would then be removed by the other tool. However, the cutters not only twist during the reciprocating motion to generate the helices but also slowly rotate in unison with the gear in order to generate the tooth contours. This generating and twisting motion results in finishing

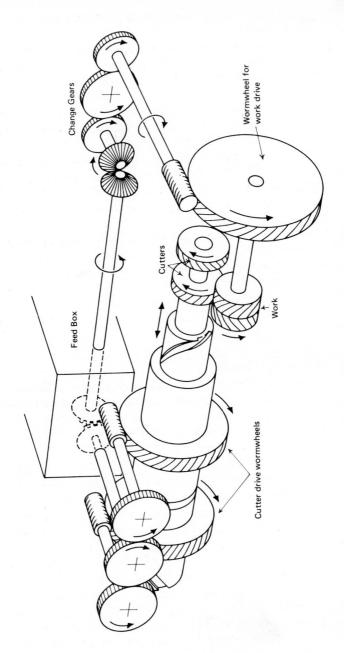

Change Gears

Wormwheel for work drive

Cutters

Feed Box

Work

Cutter drive wormwheels

Fig. 28. Basic principle of horizontal gear shaping machines

the apices of the teeth since as the cutters slowly roll out of engagement the chip produced tapers off to an infinitesimal thickness, thus cleaning out the corners perfectly.

The basic principle of the horizontal type of machine can clearly be seen from Fig. 28. Each cutter head is separately controlled by its own index wormwheel and guide assembly, the big advantage being that it enables two different gears to be processed simultaneously. The outer guide and spindle are joined together by a connecting shaft (Fig. 29) which passes through the bore of both the inner guide and spindle.

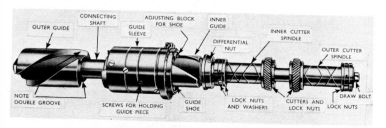

Fig. 29. Drive to cutter heads

Also the guide units are fully interchangeable with each other so that right or left hand helices can be produced by either head as required. Owing to this interchangeability it is possible to use a straight and helical guide at the same time so that a spur and helical gear can be cut at one setting. Typical of such components is the mainshaft of a car gearbox which normally has a set of helical teeth with spur synchronizing teeth adjacent to them (see Fig. 30). The two gears have different numbers of teeth and are of different diameters but providing the difference is not too large they can still be produced at one setting. The centre distance between the cutter and work spindles is constant so that the cutters are made with different diameters to suit the component. To enable the machine to produce two different numbers of teeth an attachment is fitted across the two cutter index worms so that they can index at two different rates. The machine is geared up through the index gearbox and the work spindle to a ratio to suit one cutter and its appropriate work-piece. The speed of the work spindle is now fixed but the index train across the cutter index worms enables the speed of the second cutter index to be modified to the required ratio. This enables very high pro-ductivity to be obtained since no time is wasted in lost motions — a cut is made in each direction of the cutter heads and two gears are produced in the time normally taken to produce one on a single spindle machine.

Fig. 30. Cutting spur and helical teeth at one setting

The technique of cutting two different numbers of teeth at once is known as cluster cutting. If the difference in diameters is large the technique can be modified, such that the two components are mounted at the same time, the two small gears produced at one setting and the two large gears at a different one.

Multi-cutter and multi-blank set-ups are also possible and the following examples illustrate just how universal and versatile this type of machine can be.

Multi-cutter set-ups involving two, three or four cutters are possible depending on the application. Fig. 31 shows four cutters set up to produce the teeth on starter rings for a car. Two cutters are mounted on each head on extended nose cutter spindles, the distance between the cutting edges being dictated by the face width, number of components, and the maximum acceptable overhang of the spindle out of its bearing.

Fig. 31. Multi-cutting of starter rings

Fig. 32. Cutting cluster gears

As shown each cutter is set up to stroke through a batch of four components, thus a total of sixteen components are loaded per setting. The production time is that required for a single spindle machine set up with one cutter producing four components. In this case therefore the multi cutter technique is four times as productive as a single spindle. However it must be remembered that the cutters need to be produced as a matched set of four. Fig. 32 shows a further example, a cluster gear where the two gears are generated at the same time and the components are loaded automatically.

A classic example of the multi-cutter technique, and one which shows much ingenuity, is the production of the synchronizing sleeve of a car gearbox (Fig. 33). The component has four sets of teeth, each of which has a different tooth formation in that certain teeth are removed or changed in tooth thickness. The section of teeth with the largest face width has certain teeth which are reduced in tooth thickness at one position and can be produced only from one direction. The teeth are also in alignment with each other which means that if they were cut individually each section would have separately to be aligned. Using the method shown the alignment is built into the set of cutters and once they are set on the machine they will consistently produce the correct

Fig. 33. Multi-cutting synchronizing sleeve

relationship on the component. The cutters are designed and manu-factured in sets of four — two on each cutter spindle — and they are specially located relatively to each other to give the correct axial align-ment between each bank of teeth on the component together with the correct relationship of tooth thickness. The cutters can easily and posi-tively be reset to the same relationship after sharpening, since they are designed with a suitable tenon and slot arrangement which enables them to be separated for sharpening and then relocated.

Fig. 33 also shows the finished component being unloaded from the machine, which is equipped with automatic work cycle. The component is clamped by the hydraulic tailstock which withdraws at the end of each cutting cycle for ease of unloading.

All four sections are cut at once from the solid in two and a half minutes, whereas a single spindle machine takes some twelve to fourteen minutes and requires three separate operations.

Components ideal for multi-cutter and multi-blank set-ups are shown in Fig. 34 and again are typical of synchronizing sleeves. The teeth of these components are usually such that for part of the face width only they are relieved so that there is a change in the tooth thickness. The example shown in the photograph illustrates the use of three cutters. Here two cutters are mounted on one head performing the bulk of the

Fig. 34. Multi-cutter set-up

Fig. 35. Cutting long shafts

metal removal while the third cutter on the other head performs the side trimming operation.

The horizontal machine is ideal for handling long shafts and awkwardly shaped components. Fig. 35 shows a component being produced by one cutter mounted in the standard position and the other at the rear of the head. This has the advantage that teeth widely spaced along the gear axis can be produced in one setting. The component in question is also heavy to handle so the machine has been equipped with loading rails allowing the component to be rolled into position on the fixture with the minimum of effort.

Fig. 36 shows a particularly difficult component some 200 mm in diameter and 1 metre long. In this case the component is inserted inside the main spindle of the machine. The teeth to be produced are internal and are produced from an auxiliary cutter spindle mounted across the two cutter heads. The two heads are retimed so that they relieve together and the special spindle is bolted to both heads; this extends the

Fig. 36. Cutting large tubular components

Fig. 37. Attachment for internal gears

internal capacity of the machine considerably. Internal gears can be produced also by removing the outer cutter head and mounting an extended spindle in the inner head as shown in Fig. 37.

Fig. 38. Steering gear segments

(c) Special rack-tool machines

Sykes have equipped their horizontal machines with special heads so that rack-type tools can be used for generating steering gear as in Figs. 38 and 39. This type of component is usually in segment form with tapered teeth — being segmented, it is more economical to gear shape than to hob since a great deal of time would be wasted with the latter process. Unfortunately the form produced by gear shaping with circular type cutters is not theoretically correct and the bearing produced between the mating parts is poor. The correct form can be produced only by a rack and since the components are segmented the length of the tool required is not excessive. Fast production times can be achieved by this process since one tool is mounted on each head so that cutting takes place in both directions and there is no wasted stroke. The machine is arranged with automatic reversal of feed so that a completely automatic work cycle is possible with facility from one to four cuts. The taper is obtained by inclining the work head and tailstock assemblies at the required angle.

Gear planing

The Maag machine consists of a vertical single spindle gear shaper which utilizes a rack tool instead of the conventional pinion type. The

Fig. 39. Machine for cutting steering gear segments

machine normally operates on the generating principle, although form tools operating on the slotting principle can be used.

The tool reciprocates in a vertical plane to give the required cutting action while the generating motion is obtained by combined translation and rotation of the workpiece. The action can be likened to that of a pinion rolling along a rack where the pinion represents the workpiece and the rack the cutting tool.

As it reciprocates the tool progressively cuts out one tooth space after another, the involute flanks being generated by the enveloping planes swept by the straight edges of the tool. The cut takes place during the downward stroke and the feed or generating motion is stopped during the return or upward stroke and at the same time the tool is relieved radially away from the work.

The disadvantage of this process is that for the gear to be generated in one pass the length of the cutting rack must be equal to the length of the circumference of the workpiece. For gears up to 50 mm or 75 mm diameter this is reasonable but the length of the cutting rack becomes excessive on gears over 150 mm diameter. The rack is therefore restricted in length so that it has less teeth than the gear to be cut. This means that after reaching the end of the cutting rack the gear has to be returned to its original position. This return motion is combined with indexing in that the rotary motion of the work is interrupted for a distance equal to the number of pitches cut during the generating pass. As a result continuity of indexing is not maintained so that, when returning the work, all play has to be taken up in the index drive and the generating mechanism pre-tensioned. This is usually achieved by over-travel of the tool and the generating and return motions are repeated until all the teeth in the gear have been cut. Fig. 40 shows clearly the manner in which the cut is developed and how the teeth are formed in each pass.

The Maag machines are in fact extremely versatile and a variety of work can be performed by means of suitable attachments. The rack tool can be used not only for cutting spur gears but also for right and left hand helical gears irrespective of the helix angle and number of teeth in the gear, providing the gears have the same base pitch measured normal to the helix angle.

By using suitable profiled tools, chain wheels, ratchets, splines and other special profiles can be produced by either generating or slotting. A swivelling cutter holder enables the spur type tool to be used for helical gears, or helical tools can be used without the swivel holder which enables double helical gears to be cut with a reasonably small groove between the teeth.

The planing method is very accurate for the production of the helices of gears since the angle can be set on the swivel head by sine bar pins if necessary. The tool slide moves along guide ways giving very accurate control so that the planes swept out by the cutting edges of the rack tool are straight and regular.

Racks can be produced by means of a suitable attachment but this is a slotting and not a generating process. The tool is fed progressively into depth and once the required tooth space has been produced the rack is indexed along to the next tooth and the process repeated until the required length has been reached. The number of teeth in the tool, or number of teeth produced in one pass, depends on the pitch and material. When several teeth are produced simultaneously the length of

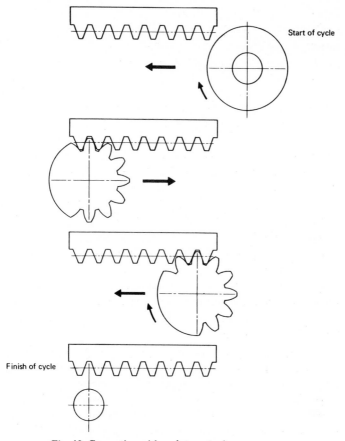

Start of cycle

Finish of cycle

Fig. 40. Generating with rack type tools

e developed chip produced can be excessive and put too large a strain
n the machine. Maag give a table which indicates the maximum num-
r of teeth which can be produced at one pass for given conditions.

Internal gears can be made by means of a special attachment which
fectively converts the cutter head from rack tools to the circular or
nion type tool. The index motion is not continuous, however, as it is
a conventional gear shaping machine — the work is indexed inter-
ittently on the return stroke of the cutter. The method is not as fast
erefore as conventional gear shaping but is a useful attachment if a
nited number of gears are to be produced. This same head which
ilizes the circular type cutter can be used for producing racks, but the

index motion is still intermittent. For pitches coarser than 10 mod. the plunge cut index method can be used for internal gears. This is really a slotting operation and a single tooth tool is used which is a male replica of the tooth space. The tool is fed progressively to depth until the tooth space has been produced and then the gear is indexed to the next tooth and the process repeated until all the teeth have been produced.

Modifications to the gear tooth can be effected in two planes

(a) along the face width of the tooth, or

(b) down the involute profile.

The profile modifications are effected quite simply by the tool in the same way as on the hob or gear shaper cutter.

The modification along the face width, known as crowning or barrelling, is effected by a small copying attachment. The cutter is radially displaced from the work during the cutting stroke by means of a simple interchangeable copying template, but the device is limited to spur gears only.

In practice helical gears of low helix angle can be produced but since the cutter is inclined to the helix angle and is cutting several teeth simultaneously it produces the correction at different points along the teeth so that the angle is extremely limited.

The planing process really comes into its own on very large pitches coarser than about 8 mod. since it is eminently suitable for large or heavy roughing cuts. The time lost in intermittent indexing is quickly made up against the continuous indexing type machine. Simple and inexpensive tools can be easily made for roughing out large gears; single tooth tools are used for gashing and these, in contrast to the rack tool do not have to be accurate. This type of tool can easily be sharpened by grinding on all edges if necessary and is easily changed and clamped in the tool holder. Adjustable two tooth tools can also be used on the very coarse pitches where a further breakdown of the developed chip is required. The tooth spaces are alternately gashed and widened by the generating method in order to reduce the roughing time as much as possible. The distance between the two cutters can be adjusted so as to change the depth of penetration and tooth thickness produced.

Other examples of the rack type generator are the Sunderland and Rollet machines, but these machines differ in that they are horizontal in configuration.

1.2.3 Rack generating

One of the big advantages of the shaping process over hobbing is its ability to generate racks. The machine must be supplied with a suitable attachment and this is usually easy to fit and remove and the conversion can be carried out very quickly.

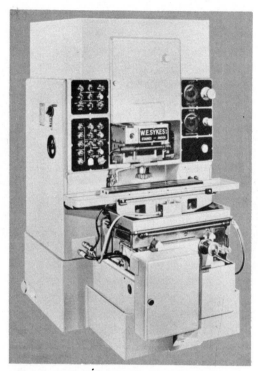

Fig. 41. Vertical gear shaper with rack attachment

Fig. 41 shows a 250 mm vertical gear shaper fitted with a rack attachment suitable for racks up to 600 mm long, although attachments are available for racks up to 1.2 metres in length maximum. For racks in excess of this length it is necessary to use a machine arranged for rack cutting only. Fig. 42 shows a machine which is capable of handling racks up to 1.8 metres in length, 6 mod. and 100 mm face width, but gears cannot be produced on this type of machine.

The rack attachment is usually operated from a master gear and rack, the gear being attached to the machine table and rotated by the table

Fig. 42. Vertical rack generator

worm and wheel through the index gear of the machine itself. Th
master rack is attached to a moving slide which carries the rack to b
machined and is in constant engagement with the master gear so that i
travels along its own length as the gear is rotated. Suitable stops an
limit switches are provided at each end of the travel of the rack gene
rated, and these reverse the table rotation or stop the main drive a
required.

The cutters used are usually made the same pitch circle diameter a
the master gear in the attachment; the index ratio of the machine i
then 1:1. For other diameters of cutter the index ratio is in proportior
to the pitch diameters of the cutter and the master gear.

One of the difficulties in cutting racks is the tendency for the racl
to distort once the teeth have been cut. Cutting the teeth on one surfac
of the component releases the inherent stresses and once the fixtur
clamps are released the component distorts. If the physical shape of th

component is suitable it should be cut *in situ* on its mounting. Unfortunately this is not always possible and therefore allowance should be made for this distortion. One possibility is to dowel the rack to its mounting before cutting the teeth and then remove the dowel and place in the gear cutting fixture, locating from the same dowel holes.

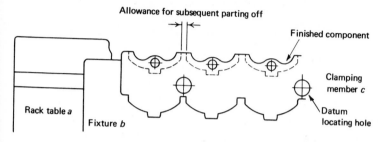

Fig. 43. Generating special form racks

The rack cutting machine or attachment is capable of cutting a variety of components provided they are in strip or rack form. Figs. 43 and 44 are an excellent illustration of what can be achieved by cutting a special form toggle in strip form and then parting off individually after the generating of the two forms. The part illustrated was originally produced by milling with two special cutters but by changing to the rack generating method the production time was reduced to one third of the milling time. The blanks are produced in strip form and suitable

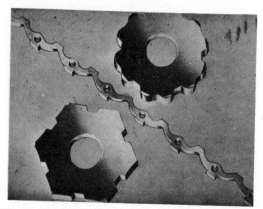

Fig. 44. Generating special form components in strip form

holes are jig drilled to serve as locations during and after the generating process. The machine is fitted with a rack cutting attachment, the table of which is indicated at *a.* On this table is mounted a support block, as used for conventional rack cutting operations, carrying fixture *b.* The fixture is arranged so that the clamping member *c* can be reversed in position on the fixture body and can then be used for holding the work for the operations on both sides of the blanks. Two blanks are loaded at one time, the position of the cutter relative to the work being set by means of a gauge. After the required profile has been machined on one side of a complete batch of blanks, the clamping member *c* is reversed on the fixture, the cutter is replaced and repositioned, and the shaping operation to produce the profile on the opposite side of a blank is performed. It should be noted that 3 mm is allowed between each of the profiles for subsequent slitting or parting off. This example is an actual case history of a job produced by C.V.A. Jigs, Moulds and Tools Ltd. (now Kearney & Trecker Ltd.) and developed in conjunction with W. E. Sykes Ltd. It was published in detail in *Machinery, 85* (August 6th, 1954).

Another typical example is that shown in Fig. 45 where the run-out at the end of the rack is restricted. This can simply be arranged by leaving out certain teeth on the cutter so that it has the same number of teeth as the job. There is obviously a limit to the length of rack that can be produced in this manner since it determines the pitch circumference

Fig. 45. Racks with restricted run-out

of the cutter and this in turn is limited, within reason, by the ram of the cutter spindle on which the cutter is mounted. Large flanges can be avoided at the ends of the teeth providing the proportions are within reason and the gap required on the cutter does not cut into its bore diameter.

1.2.4 Internal gears

The other big potential of the gear shaping process is its ability to handle all types of internal gears and special profiles, and this has many ramifications. Fig. 46 shows a typical internal gear configuration which

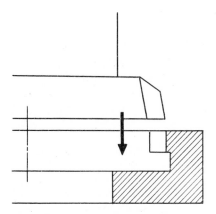

Fig. 46. Typical internal gear configuration

can only be produced by the shaping process, and the photograph, Fig. 47, an internal gear being produced on a vertical machine. There are however various types of interference which can be experienced between the cutter and gear. These are

(a) feeding-in interference;

(b) involute interference;

(c) relief interference — a combination of elements that makes the geometry such that rubbing of the cutter flanks occurs during the relief stroke of the cutter or during the infeed to depth.

Fig. 48 shows a condition peculiar to the shaping process in that the interference occurs when trying to infeed the cutter to depth. The same interference would occur if the cutter were a gear being assembled radially into the annulus. The gear would, however, run satisfactorily

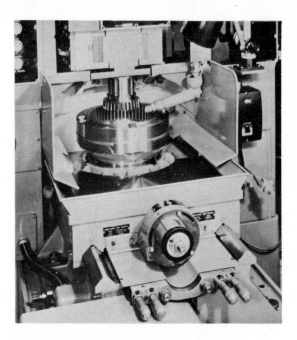

Fig. 47. Production of an internal gear

with the internal annulus if it were assembled axially at the correct centre distance. The cutter would also be satisfactory if it were physically possible to insert it at the correct centre distance.

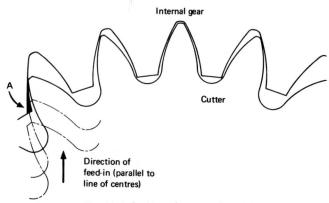

Fig. 48. Infeed interference – internal gear

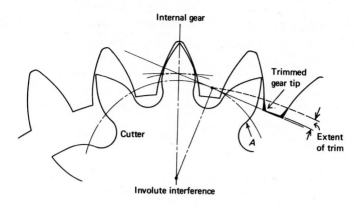

Fig. 49. Involute interference – internal gear

A further type of interference is shown in Fig. 49 and this is pure involute interference resulting from the cutter's being too small. The base circle of the cutter and the end of the line of action lie inside the bore diameter of the internal gear. Thus no involute is produced beyond this point and this portion represents a source of interference. If it is not possible to increase the diameter of the cutter the only alternative is to increase the bore diameter of the internal gear up to the point where the trim or interference starts. This will not affect the load

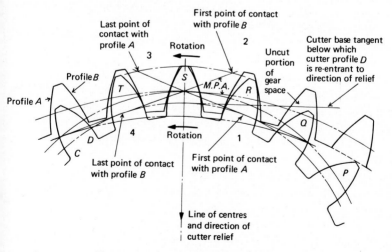

Fig. 50. Cutter generating – internal gear

carrying capacity of the gear since the trimmed portion is not in contact anyway and therefore does not contribute to the load.

The third form of interference is much more difficult to see, and indeed can occur without anyone being aware of it. The symptoms are a tendency for the cutter to pick up or rub to occur on the tooth flanks giving poor cutter life. No interference as such takes place but it is possible for the geometry to be such that certain parts of the cutter flanks are re-entrant to the line of centres. As the direction of relief of the cutter is along the line of centres no effective relief to the flank of the tool at that point can be obtained.

Fig. 50 shows a cutter in engagement at full depth with an internal gear. The direction of relief during the return stroke is assumed to be along the line of centre between cutter and work. As rotation proceeds a tooth is seen to be on the point of entering the blank at P, while Q is removing the bulk of the stock. When cutter flank C reaches point 1 it begins to produce the involute form of profile A and this is seen to have occurred in position R.

The production of profile B begins when flank D reaches point 2, and in position S both flanks are producing the final profile. A is completed in point 3 and position T shows the flank C out of engagement, while profile B is completed in point 4. It may be noted, in passing, that points 1 and 4 are on the cutter base circle diameter to give the minimum diameter at which involute action may exist. Since these points are on a diameter slightly larger than the gear bore, involute interference involving a small degree of trimming on the tip of the component would exist. This, however, has no bearing on the type of interference under discussion.

Assuming cutter relief is in the direction shown, all points on cutter flank D below a line tangent to the base circle and perpendicular to this direction will be re-entrant to the line of centre so that, on relieving and returning, the cutter tooth will foul the uncut portion of the work. Fig. 51 shows this occurring in positions equivalent to Q and R of Fig. 50.

Since the controlling characteristic is the line which is tangential to the base circle and perpendicular to the direction of relief, and the area of interference is as shown, we must (a) move the interference line or (b) remove the interference area by a roughing cut.

(a) Roughing out at reduced depth will cause the interference line to be moved in direct proportion to the reduction in depth and may remove

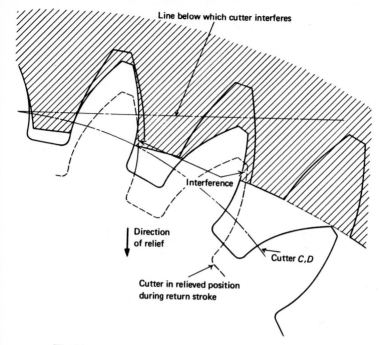

Fig. 51. Condition of relief interference when cutter relieves
along line of centres.

the interference area which exists during the finishing cut. This method
is suitable for border-line cases generally.

(b) Roughing out a tooth space having a pressure angle somewhat
higher than the finished gear will remove more stock at the bore of the
gear than at its root. Where a tip chamfer is required this may be part of
the involute formed at this higher pressure angle, thus saving the difficulties involved in producing a chamfering finishing cutter. In this case
accurate positioning of the finishing cutter is vital to ensure symmetrical
chamfers.

Finally it may be stated that this type of interference is unique to
internal gear cutting (where it occurs more often than not) and is particularly troublesome when the number of teeth in the cutter has to be
small and where the ratio of the number of teeth in the work to the
number of teeth in the cutter is less than about three. Generally some
interference may exist without unduly bad effect, and this is the reason

for its frequently passing unnoticed, particularly when the machine and tool are flexible to some extent. It may also be noted that an increase in pressure angle and reduction in depth of the component, generally possible in gear type clutches, is always of considerable assistance. For clutches 27½° or 30° P.A. should normally be possible, and such angles must be combined with a reduction of about 25% on normal depth.

1.2.5 Special applications

When generating, the direction in which the generating point, line or surface moves determines the shape produced. If this generating point is moved in varying directions while it is reciprocated then a combination of surfaces can result. If the cutter on a gear shaper were a single point tool, which reciprocated at a fixed centre distance relative to a workpiece, it would generate a cylinder, as long as the workpiece were rotated around its own axes and the axis of the cutter and workpiece were parallel. If the same two parts were mounted with the axis of the work turned through 90° so that it lay at right angles to the axis of the tool it would generate the end face of the cylinder. The tool does not necessarily have to be of the single point type — it can also be a disc. This then represents a point tool with an infinite number of points and this gives better wearing properties and the same surfaces can be generated. Simply by inclining the axis of the tool or workpiece, conical or taper surfaces can be produced.

(a) Face gears

This type of gear can be generated on a gear shaper by means of a suitable attachment which when mounted on the standard machine table turns the kinematic drive through 90° so that the axes of the work and cutter are no longer parallel but at right angles, as in Fig. 52. The cutter is of the conventional pinion type but is usually kept small in diameter and should be of the same diameter as the mating gear if possible. This type of gear is extremely restricted and the face width must be kept to a minimum since the cutter is removing active profile from the gear as it strokes toward the gear centre.

(b) Slotting

By special adjustment of the machine this type of operation can also be performed but care should be taken in observing the limitations of the process. Unlike the case of generating, the tool now cuts at all points along and around its profile, so that the length of the developed

Fig. 52. Generating face gears

chip is considerably more than would be obtained if generating an equivalent profile.

There are certain points to be observed when considering the design of the tool — care in choosing the manner in which it is presented to the work can make all the difference between success and failure. The slotting tool is mounted on the work spindle in place of the generating cutter and the index motion to the cutter spindle is isolated so that the spindle can only reciprocate and not rotate around its own axis. The work table is also arranged so that the index motion is intermittent and

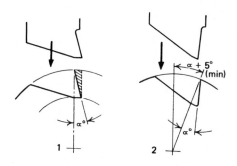

Fig. 53. Re-entrant profiles

not continuous, that is the table is stationary while slotting and indexes only when the tooth has been completed. Forms which are re-entrant relative to the line along which the tool relieves during the return stroke cannot be slotted or generated and Fig. 53 shows the reason. If the tool is designed and presented to the work in the manner shown it will remove the material drawn shaded, whereas the form could be produced if the tool were designed and presented to the work as shown in view 2.

Parallel sided splines can be generated but this process leaves a small fillet in the root of the teeth and if this is objectionable it can be removed by slotting. Certain points must, however, be observed; for example, producing the male splines can be accomplished providing the slotting tool is designed to suit the tooth space. If it is made to envelop the key as shown in Fig. 54 the tool will rub on the return stroke as

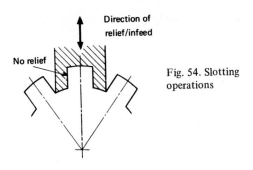

Fig. 54. Slotting operations

there is no effective relief. The tool therefore should be designed to suit the space of the spline in order to provide relief of the flanks. When considering the tool for the female the reverse applies — the tool should be designed to straddle the key in order to obtain relief. Fig. 55 shows a vertical gear shaper equipped for slotting operations, the whole cycle being fully automatic. The tool is automatically fed to depth at a given infeed per stroke and having reached the final depth of the tooth it will make one or two strokes to clean up the profile before stopping. The work then retracts slightly and the table indexes the component to the next tooth, whereupon the automatic cycle starts again and the process is repeated until all the teeth have been produced. At this point the machine will stop automatically so that the teeth on the component are not reprocessed. Sometimes multi-form tools can be used for slotting but this depends on the application and it should be remembered that making the tool with two or three teeth increases the length of the developed chip and this is the limiting factor.

Fig. 55. Gear shaping machine arranged for automatic slotting

(c) Taper cutting

Sometimes the need arises to cut gears with a slight taper, (a) because the taper is designed into the gear to allow for deflection under load, and (b) to compensate for distortion in subsequent heat treatment. The angle of taper required is usually in the order of 1° to 4° depending on the application. This can be produced in a number of ways. The best and most efficient method is on a machine equipped with an adjustable tilting table as in Fig. 56, showing the Sykes vertical machine model V10B. This machine is completely universal and can be used for conventional cutting or taper in either direction up to ± 10°.

Another method is by means of a taper attachment which is mounted on the machine table and virtually consists of an auxiliary table inclined at the required angle. This type of attachment, however, is not versatile and reduces the basic machine capacity so is suitable only for a limited range of work.

Fig. 56. Adjustable taper cutting

Other methods include inclining the machine column at the required angle — but again this restricts the machine to the production of one component only, although it is a practice quite widely used in the car industry.

One further method which can be used on tapers up to 4° is to change the cutter relief cam so that as the cutter traverses through the face width of the gear it is displaced radially to give the required taper. This method is more versatile than using attachments or inclining the column since the cams can be made so that they are fairly simple to change.

(d) Side trimming operations

Internal gear type clutches having some form of restriction at the centre which prevents the use of conventional tools or methods can be finished by a method introduced by the Fellows Gear Shaper Co. Fig. 57 shows a typical component and the cutter used to produce it.

Fig. 57. Close view of side trimming cutter and clutch

The tool has the same number of teeth as the work but is corrected or reduced on diameter so that the cutter can be inserted into the tooth space of the work after it has roughly been formed. This rough forming of the tooth spaces is usually achieved by drilling a number of holes equal to the number of teeth and large enough to allow the insertion of the cutter tooth prior to finishing. The stock is usually left on the sides of the teeth only for finishing by the trimming cutter. The tool finishes the teeth in the clutch by first cutting one side of the teeth and then reversing its direction of rotation and cutting the other side. The machine is then specially modified or adapted so that the cutter index worm cannot make a complete revolution and the degree of rotation is limited by suitable adjustable stops. By adjusting these stops the tooth thickness produced by the cutter can be controlled. The table index worm is not used; a special fixture is bolted to the top of the machine table so that the fixture can oscillate through a small arc of rotation around its own axis but cannot rotate. The relief of the cutting tool during the return stroke can then be obtained by timing the oscillation of the fixture from the cutter stroking mechanism. The operation is very fast since all the teeth are processed simultaneously and the cut is obtained by side cutting with the tool until the teeth are the required thickness.

(e) Generating special profiles

The following examples are by no means all of the types of form that can be generated, but merely serve to illustrate the versatility of the shaping process. Internal square holes can be generated by means of a suitable cutter which is considerably cheaper than a broach. Such holes can be generated with or without a radius on the corners, as the tools may be designed accordingly. Fig. 58 shows a typical example where a two tooth cutter is being used. Therefore, the machine index is set up for a 4 : 2 ratio. If sharp corners are required the cutter is designed so that the rolling or generating circle passes through the sharp corners of the work and cutter. A round hole must first be provided in a blank large enough to insert the cutter but small enough to lie inside the finished dimension across the flats of the square hole. The finished size of the hole is then obtained by setting the cutter off-centre.

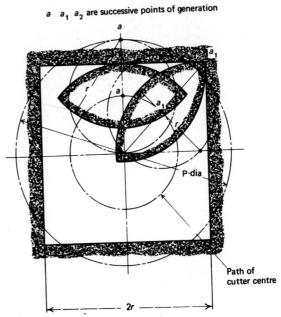

Fig. 58. Generating internal square holes

Hexagonal holes can also be generated (Fig. 59). In this case a three-sided cutter is being used, so that the machine must be geared up to an index ratio of 6 : 3. The same basic principles apply as before and a hole must first be provided to allow the cutter to be inserted.

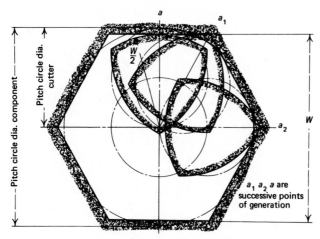

Fig. 59. Generating internal hexagonal holes

Rectangular holes are suitable for generating in this manner and this is a cheap and accurate method of production. The cutter has two sides, which are of different forms, and the machine is geared up to an

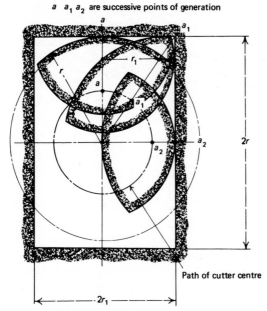

Fig. 60. Generating internal rectangular holes

index ratio of 2 : 4. The off-set is designed and built into the cutter so
that the form is off-set relative to the centre line of the cutter shank.
The form being off-set in this way means that the cutter travels a greater
distance than the length of arc on the cutter flank which generated the
long side of the rectangular hole. The difference between the length of
arc on the cutter, and the length being machined, is made up by the
skidding effect as the cutter rotates to this position. The developed
length of arc of the cutter which produces the short side of the rectangle
is equal to the length to be machined and the eccentric location does
not have any effect on the generating action (Fig. 60). The off-set makes
the cutter slide and skid on the long side of the rectangle so that it
generates a surface greater than its own developed length.

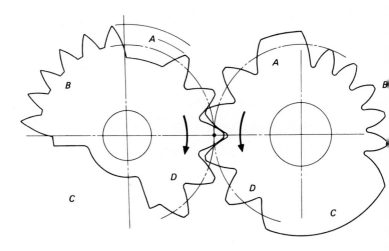

Fig. 61. Generating special double profiles

Components having two-sections of teeth which would normally be
considered as a two operation job can be generated within certain
limitations. Fig. 61 shows an excellent example. Such components are
typical of cash registers. The cutter is made with two sections of teeth
and the component is generated completely in one operation, the index
ratio being decided at the cutter design stage. This ratio is decided by
the rolling or generating circle chosen by the cutter designer and this is
the limiting factor when such components are considered.

As shown, the rolling circle is on the root diameter of the B gear, and on the mean diameter of the D gear. Therefore this fixes the generating pressure angle, the pitch and the index ratio. The main limitation of this type of generation is the degree of difference of the two forms, since the generating circle must suit both sets of teeth. When the rolling circle lies too far outside the maximum and minimum diameters of the gears the fillet generated by the tip of the cutter in the root of the teeth may be excessive or objectionable. (This is dealt with in more detail in the Section 2 Part 2, Design of shaper cutters.)

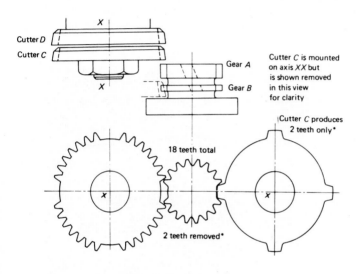

Fig. 62. Tandem cutting

Tandem cutting is another interesting application of the versatility of the shaping process. Fig. 62 shows a typical component from a gear box where two of the helical tooth spaces are projected from gear A to form gear B. As shown gear A has a total of eighteen helical teeth with two teeth removed at $180°$ intervals, and the teeth removed are in line with two tooth spaces on gear B. The two cutters C and D are manufactured as a matched pair and operate at $2 : 1$ ratio. The cutter C has four teeth and produces gear B; cutter D has twenty-eight teeth producing gear A. Cutter C strokes through gears A and B and produces two equally spaced helical tooth spaces on both, whereas cutter D produces the remaining teeth on gear *A* only. Normally this would be a two operation process involving two set-ups and double handling, so the tandem operation represents a very large production saving. The cutters must be aligned relative to each other; this can be achieved either by datum setting flats on the cutters or by a suitable setting gauge. The diameters and forms of the cutters are matched to each other when the cutters are produced and care must be exercised in sharpening to see that the same amount is removed from each cutter to maintain relationship.

Fig. 63 shows a most unusual application which involves a three cutter set-up for producing the two cams and the timing gear for an industrial engine. The two cam profiles and therefore the appropriate

Fig. 63. Special three cutter set-up

cutters are the same — merely 180° out of phase. The index ratio is determined by the gear and its appropriate cutter, as also is the centre distance of operation. This fixes the rolling circle of the cutters for the cams and the velocity ratio is obtained by the periphery of the cutter's rolling and slipping on the periphery of the cams. This in fact tends to smear out the feed marks and produces an ideal finish for the cam profiles.

(f) Gap cutting

This is the technique peculiar to the shaping generating process. It is a principle used when cutting very large batches of the same gear since the tool is virtually tailor-made to suit a given component. The infeed mechanism is not required and this is a particular advantage on Sykes type machines where the cutter is moved during the relief motion and the table remains stationary, as the table can then be locked or clamped in the operating position.

Machines which move the table to obtain the relief motion cannot, of course, be clamped during the cut and therefore cannot claim the same rigidity. Basically, if the gear is such that it can be produced in two cuts then the cutter is designed with one set of roughing and one

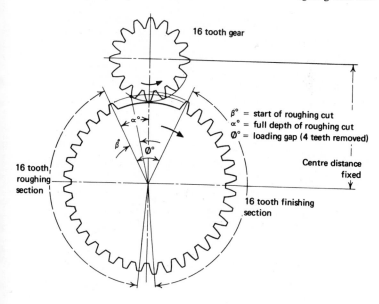

Fig. 64. Gap type cutters

set of finishing teeth, plus the loading gap. Fig. 64 illustrates the principle — the gap has to be large enough to load the gear without interference. The number of teeth in each section of the cutter is equal to the number of teeth in the gear and the centre distance between the gear and cutter is fixed. Of particular merit is the fact that roughing teeth on the cutter are only used for roughing and finishing teeth for finishing. With a conventional cutter the sharpest portions are used when roughing out and not when finishing.

The roughing teeth are reduced on depth and tooth thickness relative to the finishing teeth and great care must be exercised in the design stage to choose the right proportions since once fixed these cannot be changed without reworking the tool. A number of permutations are available for this kind of application depending on the characteristics of the gear. If the tooth depth is large and the roughing cut likely to be heavy then the initial teeth of the cutter are liable to take most of the load. This can be alleviated by making a roughing section with three or four cutter teeth suitably tapered on depth so that the cut can be applied gradually to avoid overloading. The gear is produced in one complete revolution of the cutter so that the starting position is always the same. Other permutations can be arranged; for example two load gaps can be provided in the tool (if the gear data lend themselves to it) in which case the gear is produced in half a revolution of the cutter.

Gears which require three cuts can be provided with two sets of roughing and one set of finishing teeth, the limitation being the maximum diameter of cutter the machine will accommodate. For example a 127 mm diameter gear which requires two cuts would not be a practical proposition as it would require a gap cutter of some 260 mm diameter to produce it and very few machines would accommodate such a large tool.

When this type of tool is used the machine is usually adapted with a special limit switch arrangement fitted to the guide sleeve or cutter spindle so that the main motor or feed is de-energized every revolution. This ensures that the cutter stops at the loading gap each time and the limit switch is arranged with a suitable over-rider in order that the machine can start the next cycle whilst still on the switch.

(g) Varying the helix angle

One of the basic rules of the shaping principle is that the lead of the cutter must be equal to the lead of the guide used with it. This means that a different cutter and guide are required for each helix angle to be produced. It is possible, however, with a combination of cutters, to arrange for one guide to suit different helix angles, provided that certain facts are taken into consideration and that the designer is free to accept slight deviations from the required angle.

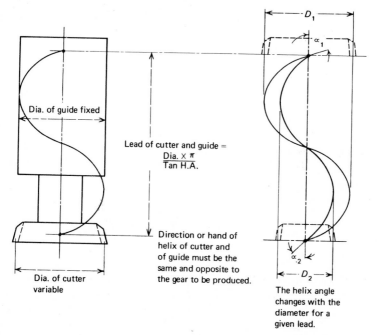

Fig. 65. Relationship of helical guide and cutter

Fig. 65 shows the relationship between the guide and cutter leads, which must obviously be the same since the guide is the means of controlling the path of the cutting edge of the cutter.

$$\text{Lead of the cutter} = \text{lead of guide}$$
$$= \frac{\text{dia.} \times \pi}{\tan \text{H.A.}}$$

The diameter of the cutter can be varied within reason, but the diameter of the guide is fixed.

If we take the case of a standard 100 mm diameter guide 30° helix angle we have:

$$\text{Lead of guide} = \frac{100 \times \pi}{\tan 30°}$$
$$= \text{lead of cutter.}$$

Since the pitch circle diameter of the cutter

$$= \frac{\text{no. of teeth}}{\text{C.D.P.}} \text{ or N.T.} \times \text{C mod.}$$

$$\text{Lead of cutter} = \frac{\text{N.T.}}{\text{C.D.P.}} \times \frac{\pi}{\tan 30°} \text{ in, or } \frac{\text{N.T.} \times \text{C mod.} \times \pi}{\tan 30°} \text{ mm}$$

Therefore tangent cutter helix angle $= \dfrac{\text{P.C. dia} \times \pi}{\text{lead}}$

$$= \frac{\text{N.T.} \times \text{C mod.} \times \pi \times \tan 30°}{100 \, \pi}$$

$$= \frac{\text{N.T.} \times \text{C mod.}}{100} \times \tan 30°.$$

(The following examples have been based on the D.P. system but the principle is the same for module.)

If the cutter were 4 C.D.P. twenty teeth it would have a pitch circle diameter of 5 in.

Therefore:

$$\text{tangent helix angle cutter} = \frac{20}{4 \times 4} \times \tan 30°$$

$$= 1.25 \tan 30°$$

$$= 35.82°.$$

If this angle is not suitable then it can be modified by changing the number of teeth in the cutter, but as this must be an integer it may be seen that the choice of angles obtained is not infinite but a series of progressions.

By providing the cutter with twenty-one teeth we obtain

$$\text{tangent helix angle cutter} = \frac{21}{4 \times 4} \times \tan 30°$$

$$= 37.15°$$

This would mean that the guide chosen in the example could only be used with cutters of 5¼ in P.C.D. if a helix angle of 37.15° were required.

If the object of the exercise were to give the maximum coverage of angles and D.P. with a given guide, and the value of the angle were not important, then further consideration can be given.

If a guide of 4 in diameter 30° helix angle were used with a 4.3 in P.C.D. cutter then a certain helix angle $\alpha_1°$ could be produced. Then all cutters of this angle $\alpha_1°$, irrespective of the D.P., could be used with this guide, providing they were 4.3 in P.C.D.

A 43T 10 C.D.P. cutter would have 4.3 in P.C.D., as also would an 86T 20 D.P. but there are very few permutations, bearing in mind that the number of teeth must be an integer.

It follows that what is required is a P.C.D. which will give as many permutations of number of teeth and D.P. as possible, for example:

6 in P.C.D. 2 D.P. 12T – 3 D.P. 18T – 4 D.P. 24T – 5 D.P. 30T – 6 D.P. 36T – 7 D.P. 42T – 8 D.P. 48T.

6½ in P.C.D. 2 D.P. 13T ——— 4 D.P. 26T ——— 6 D.P. 39T ——— 8 D.P. 52T.

From the above broad outlines it can be seen that one guide can be arranged to produce a variety of angles, providing imagination is used at the beginning and a certain flexibility is permissible.

It goes without saying that a cutter must be made for each different helix angle, for only the guide would be constant; but since the guide is the most expensive item this represents a considerable saving.

The following chart (Fig. 66) serves to illustrate what can be done with this type of approach.

Trans-verse Module	4 in (100 mm) nominal P.C.D.			6 in (150 mm) nominal P.C.D.		
	No. of teeth	Helix Angle		No. of teeth	Helix Angle	
		Deg	Min		Deg	Min
1	102	30	– 06	–	–	– –
1,25	80	29	– 36	–	–	– –
1,5	68	30	– 06	–	–	– –
1,75	58	29	– 59	87	29	– 59
2	50	29	– 36	75	29	– 36
2,25	46	30	– 28	69	30	– 28
2,5	40	29	– 36	60	29	– 36
2,75	38	30	– 42	57	30	– 42
3	34	30	– 06	51	30	– 06
3,5	30	30	– 49	45	30	– 49
4	26	30	– 35	39	30	– 35
4,5	24	31	– 32	36	31	– 32
5	22	32	– 01	33	32	– 01
5,5	18	29	– 22	27	29	– 22
6	18	31	– 32	27	31	– 32
7	16	32	– 28	24	32	– 28
8	14	32	– 28	21	32	– 28
9	–	–	– –	18	31	– 32
10	–	–	– –	18	34	– 17
11	–	–	– –	15	32	– 01
12	–	–	– –	15	34	– 17

Fig. 66. Chart showing helix angles produced from a common guide

(h) One-tooth pinions

The question can arise as to what is the largest ratio that can be obtained with a single pair of gears at parallel axes. High ratios can be obtained with a single start worm and wormwheel but they do not have the same plane of rotation and are not parallel axes in the accepted sense. If the gears have straight teeth it is doubtful if satisfactory engagement can be obtained below three or four teeth, and this would be the absolute minimum. On the other hand, if the gears are made helical it is possible to get down as low as one tooth, provided that the lead of the gear and the face width are chosen intelligently. Fig. 67 shows a one tooth pinion meshing with a 63T wheel, the details being as follows.

	Inches		Millimetres	
	Pinion	Wheel	Pinion	Wheel
Number of teeth	1	63	1	63
P.C.D.	0.508	32.004	12.903	812.552
Outside diameter	1.756	32.210	44.602	818.134
Transverse pitch	1.968 C.D.P.		12.9065 C mod.	
Face width	6		152.4	
Centre distance	16.25		412.75	
C.P.A.	22° 10 minutes		22° 10 minutes	
Ratio	63 : 1		63 : 1	

The continuity of contact between the teeth is ensured by the helical overlap. The lead or axial pitch of the pinion is equal to or less than half the face width; it cannot be more or full continuity will not be obtained. If we consider the face width as one very thin lamina, contact between the gear flanks occurs at one point only on each flank. By increasing the number the helical action is such that the point of contact between the flanks changes at each successive lamina. The point of contact moves along the helix and provided there is sufficient face width continuity of action is ensured. All phases of engagement are in simultaneous contact and therefore the transmission is smooth and positive.

The exact opposite of course is a 1 : 1 ratio at absolute minimum centre distance and this can be arranged with two one-tooth pinions. Where three one-tooth pinions are in engagement, all three being identical, having the same gear data and hand of helix, two of them revolve in one direction of rotation and the other in the opposite direction, and as the gears are in external contact this is a phenomenon of great interest.

Fig. 67. Double helical gear and one tooth pinion

1.2.6 Special machine features

(a) Infeed mechanisms

The infeed device or means of applying the cut was briefly mentioned when describing the general principles of the gear shaping process. Very little has been written on the subject, however, since most manufacturers are loath to give away 'know-how', and even less has been written on optimum infeed rates.

Most infeed devices are mechanically operated and involve 1, 2 or 3 cut camplates which can be changed or adjusted through suitable gearing. The general idea is that the gear and the tool are urged together by the cam during a given arc of rotation, and then held at that position while the gear rotates through 360°; the cycle is then repeated until the desired number of cuts have been taken.

The arc of rotation necessary to apply the cut is fixed by the cam and the 360° rotation of the gear depends also on the length of the cam periphery. This varies in detail with different manufacturers. If the table does not make 360° of rotation the cam must be lengthened or shortened as required, and some manufacturers make provision for such an adjustment.

This in itself is interesting since most people familiar with the art of gear shaping are aware of the phenomenon known as the 'dropped tooth' which usually occurs at the point where the cut finishes. The exact cause has never satisfactorily been identified but the most popular theory is that the table either has not completed 360° by the time the gear has finished, in which case a slightly thick tooth is produced, or it has made more than 360° of rotation, and owing to indexing errors or lack of stiffness an overlap of the cut has taken place resulting in a thin tooth.

Explaining such phenomena is difficult as most theories hold a grain of truth, but not necessarily the whole truth. The author's personal view is that the 'dropped tooth' is caused by a combination of factors which when in phase cause excessive 'drop' and when out of phase can cause 'drop', but sufficiently low in amplitude as to make it difficult to distinguish it from pitch or other local errors. Obviously excessive errors in the kinematic train of the shaper, particularly in the cutter and index worm wheels, can cause this condition as, if the indexing is inaccurate, overlapping of the cuts can occur and the tool starts to cut away the flanks already formed.

Apart from the phasing of the worm wheels the eccentricity or accumulative pitch error in the cutter is also important. If the 'drop tooth' were solely due to this factor it would account for the random nature of the phenomenon, since the ratio of cutter to work and the starting position of the final cut would be the critical factors. Fig. 68 shows a typical plot of the accumulative pitch errors in a shaper cutter. This can be plotted as pitch errors or eccentricity; the effect is the same and in practice it is very difficult to distinguish between the two. It is evident that if the ratio of cutter to work were a half and the final cut

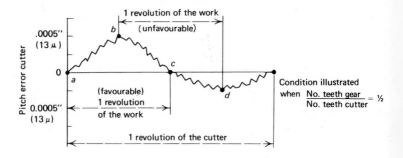

Fig. 68. Conditions favourable to 'drop tooth' on a gear

started at *a* then it would finish at *c* and the 'drop tooth' as a result of this factor would be nil. On the other hand if the cut started at *b* and finished at *d* the maximum effect would be experienced and a 'drop tooth' would be likely depending on the amplitude of the pitch errors.

Again, what at first sight would appear to be indexing errors can be caused by lack of stiffness in the machine or in the set-up. The displacement of the axes of the gear and tool to each other during the cut has the same apparent effect as indexing errors, although this can be identified by carrying out further checks for concentricity of the pitch line or root of the tooth. Unfortunately, if the machine or set-up lacks stiffness then the phase and amplitude of the 'machine flap' vary as the force levels change.

Work has been carried out at Cambridge University on behalf of W. E. Sykes into the forces involved in cutting gears by the shaping process. These forces were measured by means of a suitable dynamometer, again developed at Cambridge for this very purpose. This measures the vertical forces up the axis of the cutter spindle, the torque on the gear blank and the separating forces between the gear and the cutter. If these individual forces remain constant then the resulting force remains constant in both phase and amplitude and so does the degree of 'machine flap'. The balance of forces, however, is upset during the infeed for each cut, as can be seen from Fig. 69, which is a tracing of the cutting force measured during one of the tests at Cambridge.

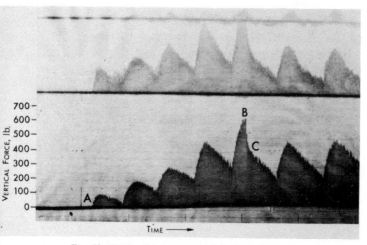

Fig. 69. Cutting forces measured during infeed

From A to B the force rises until the depth of the cut is reached then drops to the normal level at C. During the actual infeed the forces are likely to be considerably higher than the forces during the remainder of the cut. The maximum of B may coincide with a maximum or minimum of the 'once per tooth' variation of force. If the phasing is such that B corresponds to a minimum there will be little overload but if at maximum phase the force level is increased considerably.

It can be seen from the example shown that the force level during infeed increased by nearly 50%, from 450 to 650 lb. If the vertical force during infeed increased by 50% there would be an increase in the resultant force (the resultant of the vertical separating and tangential components), and as this would vary in both amplitude and phase it would be reasonable to assume that on a non-rigid machine or set-up there would be a misalignment of the tool to the work during the period of peakloading which would always occur at the 'drop out' point.

It can be seen from the above, therefore, that it is desirable to be able to control and regulate the infeed rate per stroke, and this is difficult to achieve with the cam type devices. The Sykes organization has overcome the problem by patenting a hydraulic infeed device which can be varied as required to give fast approach and withdrawal plus controlled infeed per stroke. Since the infeed per stroke can be adjusted from a simple dial it is possible to regulate the peak cutting forces while the cut is actually in progress. No precise method of determining the optimum infeed rate exists but the following notes, which were prepared by Dr. J. D. Smith and Mr. D. B. Welbourn during the work on cutting forces at Cambridge, show a realistic approach to the problem.

The heat loading on a cutter (at a given cutting speed) is proportional to the vertical force. To obtain maximum cutter life the peak vertical forces should be kept as small as possible to keep cutter heating to a minimum. An approximate analysis for normal B.S. gear teeth gives the ratio of metal removal rates due to infeed and radial feed as:

$$\frac{\pi}{2.25} \times \text{contact ratio} \times \frac{\text{infeed}}{\text{radial feed}}$$

If it is assumed that forces are proportional to metal removal rate and that the contact ratio is two then the ratio of forces due to infeed and radial feed becomes:

$$2.8 \times \frac{\text{infeed}}{\text{radial feed}}$$

Analysis of the experimental results obtained confirms that this expression for the force ratio is a good approximation to the test results for spur gears. Of course, the maximum at B may coincide with a maximum or minimum of the 'once per tooth' variation of force. If the phasing is such that B corresponds to a minimum there will be little overload.

The peak cutting force is proportional to
[radial feed + (2.8 × infeed)].

If t is the time taken to cut a gear, N is the number of teeth in the cut gear and c is the number of cuts taken, then the time is proportional to the number of strokes for infeed and radial feed. Thus:

$$t \propto \left[\left(\frac{2.25}{\text{D.P.}} \times \frac{1}{\text{infeed}} \right) + \left(\frac{\pi Nc}{\text{D.P.}} \times \frac{1}{\text{radial feed}} \right) \right]$$

Optimizing either for minimum time or for minimum force gives approximately:

$$\frac{\text{infeed}}{\text{radial feed}} \simeq \frac{1}{2} \sqrt{\frac{1}{Nc}}$$

So, if a single cut is taken with twenty-five teeth the infeed should be one tenth of the radial feed, but if three cuts are taken on a thirty-three tooth gear the infeed should only be about one twentieth of the radial feed. A higher infeed will load the cutter, whereas a lower infeed wastes time. When a 250 microns radial feed is being used the optimum infeed for a single cut twenty-five tooth gear is 25 microns per stroke. If 13 microns infeed is used there is an 11% lower peak cutter loading with an 18% longer cycle time. If 50 microns infeed is used there is a 9% reduction in cycle time but an increase of 22% in connection with the peak cutter loading.

(b) Variable speed and feed

In the past the cutting speed and feed have been determined by means of a gear box or pulleys and change gears. These give progressive steps so that certain feeds and speeds can be selected, and while this is sufficient for most purposes, occasions arise where the facility to dial any value or combination of values does have certain advantages.

Most manufacturers of gear cutting machines now provide facilities for infinitely variable feed and speed and these are generally obtained through proprietary units fitted to the machine. It is not proposed to describe these units in detail since there are so many types available and the means of obtaining the variation is not so important for this discussion as the actual use made of the facility.

The big advantage is the ease with which the optimum feed and speed can be determined for a given gear. The feed and speed rates can be changed while cutting is in progress until certain criteria have been reached. For example, the manufacturer may give the maximum rate of metal removal of which the machine is capable (although this is unusual) or he may merely indicate the maximum current rating of the main drive motor. In any event whatever the means of indicating the rate of metal removal the two variables can be dialled in until this critical rate has been reached. Maximum cutting speed is easier to determine in that this can be judged within reason by the colour of the chips or by the degree of discoloration of the tips of the tool. Once the tip of the tool discolours it is fairly safe to assume that the peak cutting speed has been reached. Maximum feed is more difficult to determine but the indications are the tendency for rubbing or metal adhesion to occur on the flanks of the tool. Unfortunately some materials have a tendency to weld more than others so that even at low feed rates rubbing is possible. In this case the number of cuts must be experimented with in conjunction with the feed rate.

The variators enable very fine adjustment to be made while cutting is in progress and the effect on the tool can be observed. Obviously the extent to which the machine labours, the accuracy produced or the deflections that take place are also factors to be considered and the feed and speed has to be adjusted with the number and depth of each cut.

The maximum rate at which metal can be removed depends on a number of factors but probably the most important is the stiffness and rigidity of the machine tool. For this reason the feed and speed are determined by the load imposed during the roughing out. The feed and speed on the finishing cut impose nowhere near the same load on the machine and it follows that there are circumstances when it would be more economical to rough out at one feed and speed and finish cut at a higher rate. This can be done on any machine whether or not it is equipped with variators or automatic change, but is not very practical on short time cycles if the operator has to effect the change by hand. Many machines can now be equipped with variators and automatic change of feed and speed which can be effected on any of the cuts. This feature can be a big advantage on gears having a long time cycle since the last cut can be taken at a higher rate. For example a gear being produced in three cuts taking twenty minutes a cut would take sixty minutes without automatic change, but if the last cut were taken at twice the speed it would only take a total of fifty minutes. If the feed and speed on the final cut were taken at twice the rate used in roughing the total time would be forty-five minutes which represents a considerable saving. Fast light cuts can usually be taken without burning the cutter — providing the increase is not too much — whereas the amount the feed can be increased is dictated by the degree of finish required.

(c) Angular relief

Higher production rates can be achieved under certain circumstances if the direction in which the cutter relieves away from the work can be varied. Fig. 70 shows an external gear and it can be seen that rubbing occurs if the cutter is relieved along the line aa but by relieving at angle α° along the line ab_1 the tendency to rub is eliminated. The angle of relief must, of course, be in the direction of the feed advance, and although the effective side relief of the cutter on the trailing flanks is increased, the relief on the leading flanks is decreased. Provided the angle of relief is less than the pressure angle, however, this does not present a problem. Again the rub (if any) occurs during the arc of approach when the bulk of the metal is being removed. During the

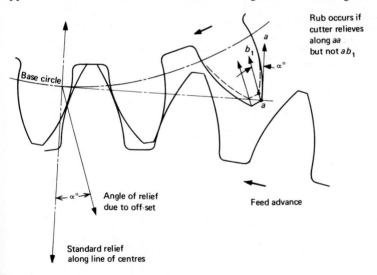

Fig. 70. Condition of relief interference when cutter relieves at angle α° to line of centres

recess action the offset works against the increased relief and the rub transfers itself to the other flank if the offset is excessive. The chances of rub occurring on the leading flank are small during the recess action, since by the time this is taking place the bulk of the tooth space has been completed and very little metal removal is taking place.

The off-set is achieved on the Fellows type machines by means of angled relief blocks which withdraw the work from the tool at an angle. The change of angle, however, on the Sykes type machine involves off-setting the saddle relative to the tool head; this is easily effected and gives infinitely variable adjustment.

Although the tool is still relieving in the same manner the effective relief is along the common line of centres and owing to the centre line of the work's being off-set the tool now relieves at an angle relative to the work.

(d) Cutter lift

Another refinement fitted to a vertical single spindle machine which offers certain advantages when producing internal gears is the automatic cutter lift. Components such as that shown in Fig. 71 are difficult to produce on conventional machines without wasting a great deal of time. Under normal circumstances it would be necessary to use excessive stroke to produce a component such as that shown in view 1 of the diagram. The cutter lift device enables the stroke to be reduced to the absolute minimum as shown in view 2, and this, of course, reduces the produc-

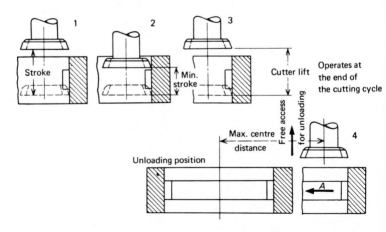

Fig. 71. Automatic cutter lift

tion time without the tool having to do any more work. In view 3 the gear has just been finished and the cutter has been lifted vertically along its own axis until the cutting edge is clear of the top of the component. At this point the saddle carrying the workpiece automatically retracts in the direction of arrow *A* taking the workpiece clear of the cutting tool to the unloading position as view 4.

The sequence just described is peculiar to the Sykes model V10B, a fully automatic and hydraulically operated machine. The cutter lift mechanism on this machine is a hydraulic device which telescopes the connecting rod between the cutter spindle and the main crank. Normally the rod is a solid member, but when the lift device is fitted a hydraulic cylinder is incorporated such that it can telescope when the cutter is to be lifted clear of the work. In the working position the hydraulic cylinder clamps the rod so that it operates as a solid member.

1.2.7 Setting up the shaping machine

As with hobbing, the accuracy built into the machine and the tool will be lost if due care is not taken at the setting-up stage. When setting up the cutter first clean the bore and locating faces on both cutter and spindle and after assembling check for true running. The same applies to the gear to be cut – it must carefully be cleaned in the bore and on the datum faces before placing in the fixture. Care taken on clocking both the gear and the cutter true is worth-while since concentricity of both members has a very big effect on the accuracy produced in the finished product. To enable the truing up to be effected quickly most modern machines are equipped with facilities for quick truing whether by power or by hand. The index change gears are usually mounted on a swing frame and the ratio is normally straightforward, depending on the number of teeth in the gear and the cutter. Care should be exercised in setting the stroke of the machine for the following two reasons.

(a) Excessive stroke for a given crank speed burns up the cutter.

(b) Incorrect setting relative to the height of the blank and fixture can cause the cutter to foul the face of the gear.

The stroke is usually 6 mm longer than the face width of the gear to be produced and has the same over-run top and bottom. It should be remembered that with crank-driven mechanisms the linear speed of the spindle is not constant and varies sinusoidally. Reference to cutting speed therefore means the peak value, which is reached half way through the stroke (the linear speed of the spindle is zero at both the top and bottom of stroke). It is necessary to provide some over-travel of the cutter relative to the blank, therefore, so that the cutter is still moving as it enters or leaves. When cutting very large face width gears it is sometimes advisable to increase this over-travel in order to avoid tearing at the edges. On these large face widths the change in cutting speed can be considerable and this can result in a change in surface finish with torn flanks where the speed of cut is too low.

Setting of the infeed varies a great deal with the infeed mechanism but when starting a new set-up two steps are necessary which are fairly common to most machines. The first is a check on whether the correct index change gears have been used — this is achieved simply by starting the main drive so that the cutter is stroking and then moving the cutter slowly into the work until it just starts to mark the gear. The gear is then

Fig. 72. Fine pitch vertical gear shaping machine

allowed to make one complete revolution and the number of marks or scratches on the periphery should correspond to the required number of teeth. The second step is partially tied to the first and is used to set the infeed. When the cutter first starts to mark the blank the infeed setting is zeroed and the appropriate depth of cut is proportioned out from this point.

One point to be checked in the operating manual of the machine is the limitation of index ratio between the gear and the cutter. For example if a 100 mm diameter cutter is used to produce a gear 6 mm diameter this involves gearing up to an index ratio of 100:6, and since not all machines are capable of being geared up so high this sort of detail should be checked.

Work holding fixtures

The same remarks apply as in hobbing — good gears cannot be cut from bad blanks or poor fixtures. Multi-loading of components can save production time but if the end faces are incorrect the resulting gears will be unsatisfactory. Comments on the type of clamping are also the same — the type of production determines the best method and long shafts should always be supported with some form of overhead support. Fig. 72 shows a fine pitch vertical gear shaper producing a shaft type of component which necessitates some form of overhead support. This is worthy of special note since the construction of the support gives optimum stiffness and accuracy and yet is universal and easy to unload.

1.2.8 Calculating the shaping time

This is a simple matter once the basic mechanics of the operation are understood. The determination of the cutting speed is straightforward and the value depends on the type of material to be cut. Again it is not possible to generalize since the speed is influenced by so many factors but the table on page 53 gives a rough indication for general purposes. Having arrived at a suitable value and knowing the face width to be produced, the stroke and hence the number of strokes per minute can be determined. The cutting speed in feet per minute is equal to the linear speed of the ram and for a given stroke S and strokes per minute N the speed equals $\dfrac{S}{12} \times \pi \times N$ ft/min $\left(\dfrac{S}{1000} \times \pi \times N \text{ metres/min} \right)$.

The most difficult factors to fix are the number of cuts and the feed per stroke, and as very few of the machine tool manufacturers give the output of their machines in terms of cubic inches of metal removed per minute experience has to be the guide. Again the factors which affect the choice are wide and varied. The rigidity of the machine is the critical factor and apart from this the configuration of the component, the rigidity of the fixture and the nature of the material all play their parts.

Machines having a capacity up to 250 mm diameter 6 mod. should generally be capable of achieving, on spur gears, rates as in the following table (Fig. 73):

Material		Free cutting		Tensile strength 80 kg/mm^2	
Mod.	D.P.	Cuts	Feed	Cuts	Feed
4–6	4–6 inc.	3	0.010 in 0.25 mm	3	0.008 in 0.20 mm
3	8	3	0.012 in 0.30 mm	3	0.010 in 0.25 mm
2–2½	10–12 inc.	2	0.012 in 0.30 mm	2	0.010 in 0.25 mm
1¼ –1¾	14–20 inc.	2	0.010 in 0.25 mm	2	0.008 in 0.20 mm

Fig. 73. Table of feeds and speeds

The rate of radial feed, the number of cuts and the cutting speed having now been obtained, the only variable is the rate of infeed. On machines using mechanical or cam type infeed mechanisms there is usually little facility for changing the rate unless the cam itself is changed. An indication of the rate of infeed is however usually given by the manufacturer. Otherwise, where facility for adjustment of the rate of infeed is given, the best average value to use is 25 microns per stroke. Rates in excess of this figure, particularly on hard materials and coarse pitches, tend to place an unnecessary load on the tips of the tool.

The radial feed per stroke of the cutter is the amount that the cutter advances around its own periphery during one stroke, and this measurement is usually made on the pitch circle diameter. It follows, therefore, that in one revolution of the cutter it makes:

$$\frac{\text{pitch diameter} \times \pi}{\text{feed}} \text{ strokes.}$$

If the speed of reciprocation of the cutter were N rev/min, it would take $\dfrac{\text{P. dia.} \times \pi}{\text{feed} \times N}$ minutes to make one revolution.

Assuming the number of teeth in the gear to be T_g and the number in the cutter to be T_c then in one revolution of the gear the cutter would make:

$$\frac{T_g}{T_c} \text{ revs.}$$

It follows that the time taken for the gear to make one revolution is:

$$\frac{\text{P. dia.} \times \pi}{\text{feed} \times N} \times \frac{T_g}{T_c} \text{ minutes.}$$

If the gear is being produced in C cuts then, ignoring the infeed time, it would make C revolutions.

Therefore the time would be

$$\frac{\text{P. dia.} \times \pi}{\text{feed} \times N} \times \frac{T_g}{T_c} \times C \text{ minutes.}$$

Assuming the depth of tooth to be D and the infeed rate B per stroke then the time taken to infeed to depth would be:

$$\frac{D}{B} \text{ strokes at } N \text{ rev/min} = \frac{D}{B} \times \frac{1}{N} \text{ minutes.}$$

Since it is usual to allow a small approach distance for the cutter when feeding to depth, this should also be taken into account; thus if

$$A = \text{approach distance}$$

then
$$\frac{(D + A)}{B} \times \frac{1}{N} \text{ minutes.}$$

is the time taken to infeed.

The total time taken to produce the gear is:

$$\left(\frac{\text{P. dia.} \times \pi}{\text{feed} \times N} \times \frac{T_g \times C}{T_c} \right) + \left(\frac{(D+A)}{B} \times \frac{1}{N} \right) \text{minutes.}$$

The above notes regarding calculation of production times apply to all types of gear but when cutting helical gears certain other factors should be borne in mind. For instance, the cutting speed is the rate at which the point of the cutter moves across the workpiece and with helical cutting two components of speed have to be considered. The speed down the axis of the cutter spindle is the same as for spur gears but owing to the helix angle the path travelled by the tips of the cutter is longer than the stroke since it travels down the developed helix, with the result that the cutting speed along the helix is the cutting speed along the axis divided by the cosine of the helix angle.

The number of strokes per minute for helical gears should therefore be reduced to compensate for this factor. Similarly the number of cuts and radial feed should be different when cutting helicals and the values given in Fig. 73 should be reduced slightly depending on the helix angle.

Racks

When cutting racks the same basic formulae apply such as time per revolution of the cutter

$$= \frac{\text{P. dia.} \times \pi}{\text{feed} \times N} \text{ minutes.}$$

If the length of the rack to be produced is L then the cutter makes

$$\frac{L}{\text{P. dia.} \times \pi} \text{ revolutions.}$$

Then to produce length L, time required

$$= \frac{\text{P. dia.} \times \pi}{\text{feed} \times N} \times \frac{L}{\text{P. dia.} \times \pi}$$

$$= \frac{L}{\text{feed} \times N} \text{ minutes.}$$

Therefore the total time assuming C cuts and $\dfrac{(D+A)}{B} \times \dfrac{1}{N}$ infeed

$$\left(\frac{L}{\text{feed} \times N} \times C \right) + \left(\frac{(D+A)}{B} \times \frac{1}{N} \right) \text{ minutes.}$$

The gear cutting operation sheet should be as short and explicit as possible so that the operator has no need to wade through a wealth of data which are of no interest to him in order to find the information essential to cutting the gear. The sheet should include

(a) brief gear data to enable the gear to be identified at any time;

(b) measuring data to enable the gear to be checked and also machine settings for inspection equipment;

(c) outline sketch of the gear showing location surfaces;

(d) fixture data and datum clocking registers;

(e) cutting tool data and gauges if required;
(f) type of machine, feed, number of cuts, speed, stroke, change
    gears and depth of each cut;
(g) special notes — helical guide data if required;
(h) degree of accuracy required;
(i) internal reference numbers.

Such data are sufficient for all general purposes.

1.2.9 Special shaping processes

a) Shear speed

A very unusual application of the gear cutting process is the 'Shear Speed' machine (Fig. 76), patented and registered by the Michigan Tool Co., U.S.A.

The process is similar to broaching and as such does not suffer from some of the limitations of the generating process. On the other hand it does have its own limitations. It is a high production method of producing gear teeth and similar parts which is ideal for mass production of a given component, although it is by no means limited in this respect.

All the teeth in the part are produced simultaneously in the shear speed process. The cutting tools, equal in number to the teeth in the part, are assembled radially in the tool holder. The size and shape of the tools depend on the shape of the tooth space of the component and each tooth space need not necessarily be the same. Again the tooth spacing can be irregular if required as the cutting tools are inverse replicas of the teeth to be produced and are manufactured individually and set in the holder. The tools themselves are form relieved behind the cutting edges in order that the correct form may be maintained after sharpening which can be carried out on a surface grinder. All teeth can be ground at one setting and each set of tools is normally supplied to produce a specific part.

The tools are assembled into a tool head (Fig. 74) which remains stationary in operation, apart from relieving the cutting tools away from the component during the return stroke. The component itself is placed on a suitable work holding fixture which is reciprocated past the cutting tools at the required surface speed by means of a large hydraulic ram. The tool head assembly is an ingenious mechanism consisting of three main elements:

(a) a radial member which controls tool alignment;

Fig. 74. Tool head of 'Shear Speed' machine

(b) a double cone shaped unit which feeds all the tools in at the same time; and

(c) a retainer housing.

The gear blank is reciprocated vertically past the cutting tools while all the tools are fed in simultaneously a given amount per stroke and automatically retracted during the return stroke to prevent drag over the cutting edges (see Fig. 75).

The rate of feed per stroke is controlled by a cam and ratchet mechanism which can be varied for different feed rates. The total infeed distance of the tools is equal to the depth of the part to be produced plus some 0.38 mm initial starting clearance between the tool and the work. Average feed rates vary between 25 microns and 100 microns per stroke and the cutting speed is some 6 metres per minute. These figures obviously vary with the material to be produced but are good average values, and as in the gear shaping process the length of stroke is adjustable and 3 mm overtravel is required at each end of the stroke. At the completion of the cutting cycle the tool head lifts clear of the work for easy unloading, which takes roughly 8 to 10 seconds, to which must

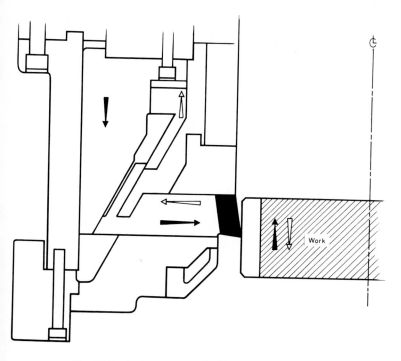

Fig. 75. Producing a gear on the 'Shear Speed' machine

added 10 to 15 seconds to unload the work. The actual cutting time is extremely short, for example:

(a) 12 D.P.   13T Spur Gear   4.16 mm deep, 19 mm face width, 32.6 mm diameter = 34 seconds each = 96 per hour at 100% efficiency

(b) component 63 mm face width, 4.68 mm depth of tooth, material SAE.8620 = 68 seconds to cut = 86 seconds floor to floor = 42 components per hour at 100% efficiency.

This shows the typical advantage in machining time with the broaching method over conventional machining/generating techniques, provided the limitations and high tool costs are acceptable and can be amortized by the high volume production.

The shear speed process is applicable to all forms having straight teeth, such as sprockets, gears, special forms, splines, ratchets and serrations, and these can be symmetrical or asymmetrical with equal or unequal tooth division. Helical teeth cannot be produced and although some internal gears can be cut there are certain limitations, particularly on the minimum diameter. Parts close to a shoulder can be produced and in this respect the limitations are roughly the same as for the conventional gear shaping process. An idea of the size of gear that can be produced can be gained from the following chart.

Model	1833	3053	3073	18105	30736	30206
Maximum diameter	3 in 75 mm	5 in 125 mm	7 in 175 mm	10 in 250 mm	13 in 330 mm	20 in 500 mm
Minimum diameter	1 in 25 mm	3 in 75 mm	5 in 125 mm	7 in 175 mm	6 in 150 mm	10 in 250 mm
Maximum D.P.	8 3 mod	6 4 mod.	6 4 mod.	3 8 mod.	1.9 13 mod.	1.9 13 mod.
Minimum D.P.	16 1½ mod.	16 1½ mod.	16 1½ mod.	16 1½ mod.	12 2 mod.	12 2 mod.
Maximum face	2¾ in 70 mm	2¾ in 70 mm	2¾ in 70 mm	4¾ in 120 mm	6 in 150 mm	6 in 150 mm
Main motor (h.p.)	15	15	25	40	75	100

Fig. 76. Michigan 'Shear Speed' machine

b) Gear skiving

This is a relatively new process for the generation of internal spur gears which has a number of limitations, and is suitable only for large quantities of the same component. The process is a combination of known techniques which when combined enable internal spur gears to be produced faster than by the conventional shaping method. The technique of crossing the axes to improve the cutting action has been known for many years and is dealt with in detail later in this book (see Part 1.3.1, Gear shaving).

The cutting tool in this case is a gear shaper cutter having a helix angle equal to the required crossed axis angle and with cutting edges normal to the helix. When the cutter is engaged with the gear and the axes are inclined to each other two directions of movement are experienced as they roll together. First is a sliding motion between the involute flanks which increases towards the tips of the teeth. A secondary motion is then experienced owing to the crossing of the axes and this is in a direction parallel with the axis of the teeth. The effect of this latter component becomes more severe as the peripheral speed of the gears increases and is proportional to the speed and crossed axes angle.

This sliding movement along the teeth is used to advantage to give a cutting action and by providing a positive drive between the work and the tool the axial feed enables the cutting edges to sweep out the tooth spaces. If reference is made to Fig. 77 the relative components of sliding

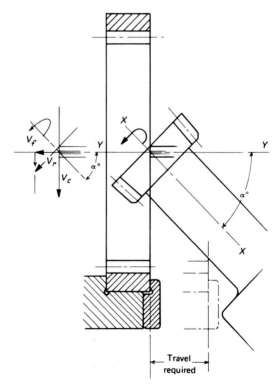

Fig. 77. Principle of gear skiving process

and cutting can clearly be seen. The cutting tool rotates around the axis XX and gives a peripheral speed component V_r and a cutting velocity V_c around axis YY. The remaining component V_f is the velocity of feed in the direction of the axis of the component tooth and it is obvious that tooth velocity V_r and V_c is dependent on the speed of rotation around axis XX and the crossed axes angle.

It can be seen that crossing the axes has many advantages but it also introduces a basic difficulty in that there is only one point where the cutter produces a true involute profile and all points produced in front of and behind this position are not involutes. The cutter profile must therefore be corrected so that the overall form produced is correct. The amount of correction applied to the cutter depends on the characteristics of the cutter and gear, the crossed axes angle and the length of the line of action and is therefore correct only for one application.

Owing to this severe limitation the cutter is limited to components within a given tooth range and therefore is economic only for large batch production. There is a further limiting factor which is purely physical, as can be seen from Fig. 77. The spindle on which the cutter is mounted has to be of such a length and diameter as to let the cutter feed through the face width without fouling the outer diameter of the component or the fixture.

If the teeth are let down inside the top face of the component this imposes even further limitations and the maximum obtainable axial transverse is dependent on the cutter diameter and the crossed axes

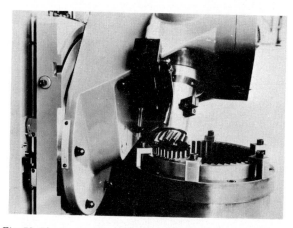

Fig. 78. Pfauter machine with special cutting head for gear skiving

angle. It is advisable, therefore, to assess thoroughly the suitability of the components for production by this process by making a simple layout as shown in Fig. 77.

Perhaps the biggest advantage of this method is the fact that unlike the gear shaping process no problems arise from the reversal of masses so there are no inertia effects to overcome, and it is not necessary to relieve the tool from the work on each stroke. The cutter rotates at speed around its own axis in the same manner as a hob and indeed it virtually is a hob with a number of starts equal to the teeth in the cutter but with only one tooth in each start. The system is being developed commercially by Pfauter in Germany and Fig. 78 shows their P630 machine equipped with the special cutting head; this indicates in detail the physical limitations of the process.

PART 1.3: FINISHING PROCESSES

1.3.1 Gear shaving

The modern gear generator has been developed to a remarkably high degree of accuracy and yet difficulty is still experienced in obtaining the quality required by present day standards. It is still necessary therefore for further finishing operations to be carried out to refine the errors and the fastest and most economical method is gear shaving.

The process consists of using a rotary tool in the form of a gear having a helix angle different from that of the gear to be produced. Cutting edges are provided in the form of a series of slots running vertically down the flanks from tip to root. These cutting edges sweep over the gear flanks and remove fine hair like chips when the tool and gear are rotated in tight mesh. The operation is a cutting and not a cold working process and to obtain a free cutting action with a minimum of pressure the axes of tool and work are crossed. The number of teeth in engagement at any one instant should be as large as possible since the tool drives the work solely by contact of the teeth, there being no direct drive between the respective spindles.

The gear is mounted on live centres attached to a slide which reciprocates the gear back and forth across the face of the tool. The speed of rotation and rate of traverse can be varied to obtain optimum cutting conditions.

(a) Principle of the process

When two gears engage pure rolling takes place at the pitch line only and at any other point along the line of action the contact between the flanks is a combination of sliding and rolling. This is clearly illustrated in Fig. 79 which shows the linear variation of the sliding velocity of the tool profile P_t in relation to the gear profile P_g along the line of action AB.

This relative sliding can be found by multiplying the sum of the angular velocities by the distance from the contact point to the pitch point measured along the line of action.

Relative sliding at B

$$= V_1$$
$$= (\omega_{rg} + \omega_{rt})\, BI.$$

The sliding is proportional to the distance along the line of action from the pitch point, thus at the pitch point the distance BI is zero and

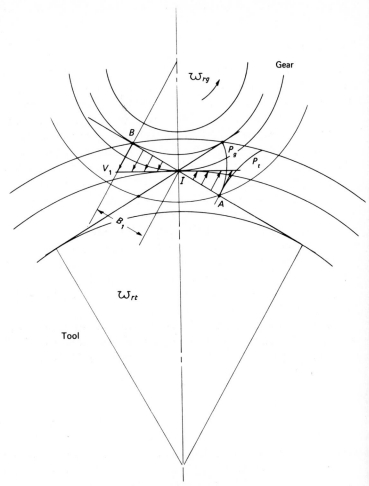

Fig. 79. Sliding along the involute

the sliding VI is also zero, i.e. pure rolling only takes place. The sliding along the line of action AI takes place in the opposite direction as can be seen from Fig. 79.

This sliding is one of the contributory factors toward making the tool cut but is not sufficient on its own since the cutting slots on the tool run in the direction of sliding. There must therefore be some displacement of the profiles of the tool and the gear across or at some angle to the cutting edges. This is done by crossing the axes of the two members thus setting up a further component of sliding axially along the teeth and across the cutting edges.

This component can be clearly seen by considering two helical racks (Fig. 80) meshing with their axes or planes of movement at right angles to each other. If the rack A were moved with a velocity V_a in the direction shown it would cause rack B to move with a velocity V_b in a direction at right angles to it. Obviously, to transfer this motion from one direction to another the flanks of the racks must slide over each other at a resultant velocity V_R. If the teeth of rack A were provided with cutting edges down the flank then providing pressure were exerted between the two members it would act in a similar manner to a surface broach and remove metal from rack B.

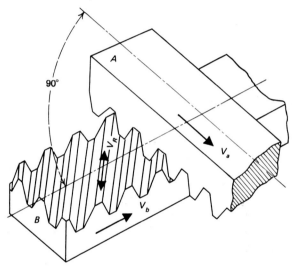

Fig. 80. Sliding along the flanks

If we now consider the extreme case where $\alpha = 0°$ and the axes are parallel and not crossed, it can be seen that the relative sliding along the flanks would be zero. Obviously, therefore, the best cutting conditions are obtained when the crossed axes angle is high; unfortunately there are other considerations which dictate the best value. If we now consider the tool and gear at crossed axes angle $\alpha°$ and no longer having infinite diameter (a rack) then we have the condition shown in Fig. 81.

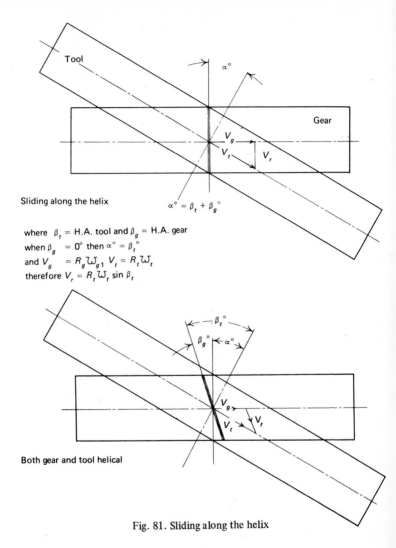

Sliding along the helix

$$\alpha° = \beta_t + \beta_g°$$

where β_t = H.A. tool and β_g = H.A. gear
when β_g = 0° then $\alpha° = \beta_t°$
and $V_g = R_g \omega_{g1}$, $V_t = R_t \omega_t$
therefore $V_r = R_t \omega_t \sin \beta_t$

Both gear and tool helical

Fig. 81. Sliding along the helix

The velocity of a contact point on the gear

$$V_g = R_g \, \omega_g$$

where R_g is the pitch circle radius of the gear, and ω_g is the angular velocity of the gear.

The velocity of the equivalent point on the tool

$$V_t = R_t \, \omega_t$$

(when the helix angle of the gear $\beta_g = 0°$).

It can be seen therefore the resulting speed of slip along the flanks
$$V_r = \omega_t\, R_t\, \sin \beta_t.$$
The crossed axis angle $\alpha = \beta_t \pm \beta_g$ and when both gear and tool are helical the situation is more complex and the speed of sliding becomes
$$V_r = \omega_t R_t\, (\sin \beta_t + \cos \beta_t\, \tan \beta_g)$$
To summarize, therefore, we find that if two helical gears are engaged together with axes crossed at angle α° and rotated at an angular velocity ω_r we have (a) a component of sliding down the flanks V_1, (b) a component of sliding along the flanks V_r.

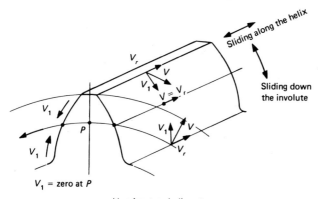

V varies at each diameter.

Fig. 82. Speed and direction of cut

It is obvious from Fig. 82 that this gives a further resultant velocity V diagonally across the flanks which varies both in magnitude and direction at different radii. The change of direction occurs at the pitch line where the involute sliding is zero but the component of sliding V_r along the flanks is still taking place.

It has already been pointed out that the cutting edges of the tool run vertically down the flanks from tip to root, therefore if we consider the contact at the pitch line the direction of cut $V = V_r$ and this is at right angles to the cutting edge of the tool. The velocity of cut V varies in intensity and direction with the height of the tooth and thus the angle relative to the cutting edge of the tool varies. This is clearly illustrated in Fig. 83. Different conditions of metal removal therefore result at successive positions down the flanks. It is highly probable that the trouble experienced in shaving tool manufacture in producing accurate involute profiles on pinions with low numbers of teeth stems from this variation.

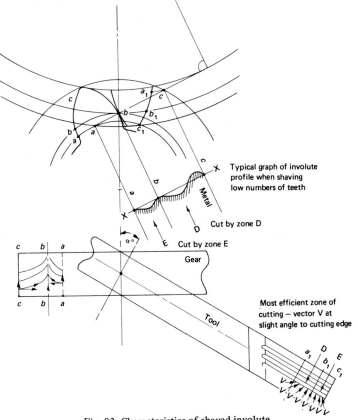

Fig. 83. Characteristics of shaved involute

Anyone familiar with this condition will recognize the characteristics of the involute shown on the graph in Fig. 83.

It would appear from the diagrams of the vector speed of cut that the best conditions for metal removal are just below and above the pitch line, i.e. because of the angle of cut relative to the slots in the tool. This would therefore account for the well known characteristic obtained under these conditions.

From the foregoing notes it can be seen that the velocity of cut can be controlled by varying the angular velocity of the tool and by changing the value of the crossed axes angle. There are however limitations regarding the value of the crossed axes angle since this also determines the contact pressure between the teeth and the length of the line of

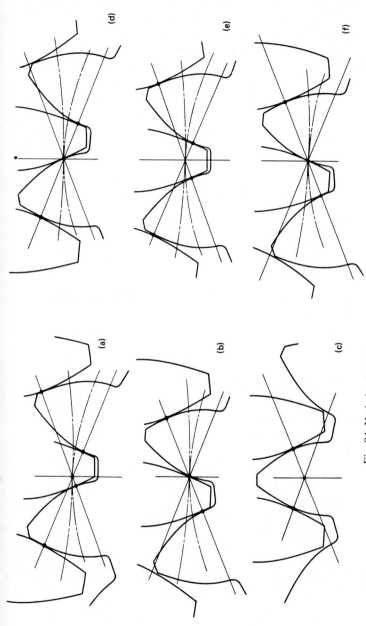

Fig. 84. Variation in contact points

action. The importance of these two items is obvious since the contact pressure determines the rate of metal removal and the line of action the control that the tool exercises over the gear while cutting.

The contact pressure is determined by two factors (a) the number of simultaneous contact points (b) the crossed axes angle.

The variation in contact points between the teeth can best be seen from Fig. 84 which shows successive phases of engagement between a gear and a tool. At (a) and (e) the teeth are in contact at four points, at (b), (d) and (f) in contact at three points, and at (c) contact occurs only at two points. The contact pressure is determined by the depth of cut and at point (a) the contact per square inch is at a minimum since it is shared at four positions, whilst at (c) it is at a maximum since only two points share the load.

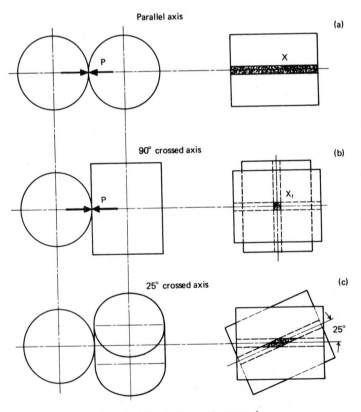

Fig. 85. Effect of crossed axes angle

The rate of metal removal is therefore at a maximum in position (c), i.e. in the vicinity of the pitch line which has already been shown to have the most favourable position from the point of view of efficiency of cut.

The effect of crossed axes angle on contact pressure can be easily shown by Fig. 85 which has been used many times before by various authors.

When the crossed axes angle is zero the axes are parallel and the zone of contact achieved by pushing two cylinders together under pressure P is X. If the crossed axes angle is 90° and the contact pressure P the same the contact area is reduced to X_1.

It follows, therefore, that the higher the crossed axes angle the greater the unit pressure obtained and the greater the rate of removal of metal. The best value for the crossed axes angle is about 15° and this has been found from practical trials, although there are certain applications where lower values must be used, these being discussed later.

(b) Rotary and rack type tools

There are two basic kinds of tool, the rack type and the rotary type. Although the former is now virtually obsolete it was one of the first production machines and the gears were processed by rolling them in tight mesh with a rack type of tool. The rack was reciprocated back and forth and fed into depth by the required amount at the end of each stroke. The racks were usually made up of individual blades which were assembled together in a suitable holder. This had the advantage that if a tooth were damaged it could easily be replaced. Each blade had serrated edges running from tip to root in a similar manner to the rotary type. The limitations of the rack process lay in the cost of producing the tools and the fact that the maximum diameter gear was limited to approximately 150 mm. The length of the rack must obviously be greater than the circumference of the gear being shaved; therefore to keep the rack length to reasonable proportions the diameter of the gear was limited.

Both the rack and rotary type tools are made in high speed steel and are capable of being resharpened some five or six times with a tool life between sharpenings varying between 5 000 and 15 000 gears.

The serrations down the flanks which form the cutting edges terminate in drilled holes in the root which serve a dual purpose since they allow clearance for the tool when producing the serrations and also enable the fine swarf to be washed away easily by the cutting oil.

(c) Techniques of shaving

There are a number of variations of the gear shaving process available each adapted to suit the conditions peculiar to some branch of the engineering industry. There are two basic methods as distinct from techniques — these are the radial and tangential methods. The radial method has already been discussed briefly; it involves bringing the tool and gear into tight mesh radially by reducing the centre distance at each end of the pass and processing both flanks at the same time. The tangential method involves selective shaving or processing one flank at a time by applying a brake to either the tool or the gear. This resistance couple is applied so that the contact pressure between the tool and gear takes place on one flank only.

The operation is repeated in reverse to shave the other flank and this method is used extensively for shaving large turbine gears since, by varying the couple, the bearing may be modified as required. It is very useful for shaving large face widths which have undulations or waves along the teeth since it enables better control to be obtained over the high spots.

When using the radial method the tool is sometimes serrated on one flank only leaving one face without serrations; this enables selective shaving of one flank of the gear to be achieved.

Fig. 86. Zone of metal removal

Showing zone of metal removal without axial displacement of the tool relative to the gear

If the tool and the gear were engaged radially in tight mesh under pressure the zone of metal removal would be as Fig. 86 as long as the gear and tool were not displaced relative to each other. The zone of metal removal is quite extensive, particularly at low crossed axes and on narrow face width gears the impression can sometimes be gained that the whole face width has been processed. This is not so however and there must be relative displacement between the axes for the gear to be completely finished.

It is the nature of this displacement that determines the shaving technique and it is unimportant whether the tool or gear is moved.

Transverse method

This was the earliest method and probably the simplest since only one feed slide is required and the multi-slide used on the other techniques is not essential. The work and tool are set to the crossed axes angle and one or the other is traversed along the axis of the gear. At each end of the stroke the centre distance between gear and tool is reduced and the direction of rotation reversed so that a cut is taken on each stroke. If required the last one or two strokes can be taken without a cut increment therefore burnishing or spring cuts are made finally to finish the gear. It can be seen from Fig. 87 that all the work in this method is carried out by one point on the tool and the wear is not distributed.

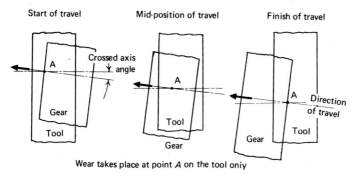

Fig. 87. Transverse shaving

The tool can be moved along its own axis however by varying the spacers on the arbor and this has the effect of bringing a fresh part of the tool into contact with the work. It also changes the pivot point, and although this is permissible on open type gears it cannot always be applied on restricted shoulder gears (see notes on shaving shoulder gears, page 144). This technique is particularly suitable for shaving very wide face width gears, long involute splines and internal gears.

Underpass or 90° shaving

In this technique the direction of travel of the tool relative to the gear is in a plane at right angles to the work axis (Fig. 88). This is the fastest method of shaving since owing to the direction of feed motion the stroke required to process the entire face width of the work is the

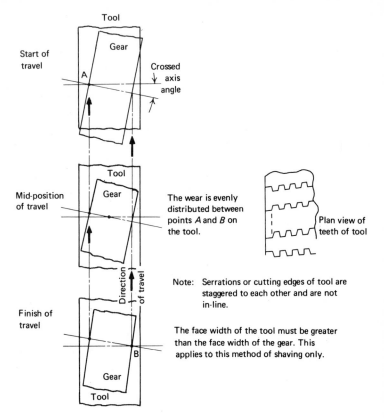

Fig. 88. Underpass method of shaving

shortest that can be obtained. When originally introduced it was intended for single pass shaving, that is to say, the gear was processed to size in one cycle — one pass forward and one back without any further upfeed or reduction in centre distance. In practice, however, the technique is now used in a similar manner to the other techniques and multi-passes are taken with upfeed between each stroke. This is mainly due to the failure of the user to keep the required shaving stock to the absolute minimum, if single cycle shaving is to be effective the shaving stock should not exceed 150 microns on diameter over pins or 75 microns on centre distance against a sizing master. To remove more stock than this really requires a number of incremental passes. When this technique is used there are two pre-requisites — (a) the face width of the tool must be larger than that of the gear and (b) the serrations in the tool must NOT be in line or annular relative to the tool axis, but must be staggered with each other with a differential action. It follows, therefore, that this technique is NOT suitable for wide faced components but is ideal for components with restricted shoulders since there is no relative displacement of the faces of the tool and gear to each other. The clearance between the tool and the interference point on the shoulder is neither diminished nor increased and therefore overtravel cannot give cause for alarm. Again the wear on the tool is averaged out across its entire face width since every part of the tool comes into engagement with the gear as it is fed across. Two work slides are necessary, however, one to set the C.A.A. between the tool and gear and the other to set the required direction of feed movement.

Diagonal or traverpass

In common with the last technique two slides are required, although in this case the direction of feed motion can be any angle between the two extremes (see Fig. 89). It would be fair to say that this is probably the most widely used technique today since it combines the best average conditions of the preceding techniques. It has the advantage of the transverse technique in that special tools are not necessary, the face width need not be larger than the gear and differential serrations are not essential. It has the advantage of the 90° method in that the whole face width of the tool is used, thus averaging out the wear. The traverse angle is the angle between the direction of traverse and the work gear

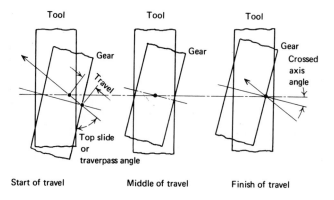

Fig. 89. Traverpass method

axis and is determined as follows

$$\text{maximum angle of traverse} = \tan^{-1}\left[\frac{\text{tool face} \times \sin \alpha^{\circ}}{\text{gear face} - (\text{tool face} \times \cos \alpha^{\circ})}\right].$$

This angle can be set at any value between $0^{\circ} - 90^{\circ}$ but the best conditions are achieved between $30^{\circ} - 60^{\circ}$.

Again single or multi-pass shaving can be utilized but owing to the difficulty of controlling shaving stock under mass production conditions, it is more usual to use multi-pass with upfeed increments at the end of each stroke.

(d) Crowning

One of the most prolific causes of failure in gears is misalignment and the elastic characteristics of the mountings apart from the errors in the gears themselves.

Misalignment can result in end bearing and therefore the stresses are concentrated at the ends of the teeth where they are most vulnerable.

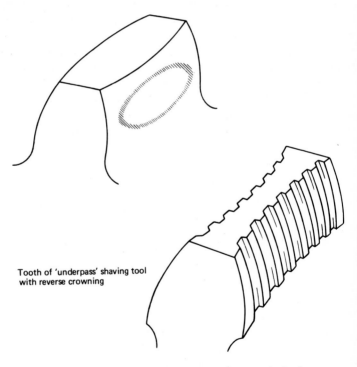

Tooth of 'underpass' shaving tool
with reverse crowning

Fig. 90. Typical crowned tooth showing bearing obtained

By adapting the crowned tooth (Fig. 90) much of this end bearing can be overcome without reducing the load carrying capacity to any great extent.

When crowning with the tool the underpass method only can be used and the tool itself must be wider than the gear to be shaved and the cutting slots down the tool flanks must be staggered with each other so that consecutive teeth are not in line. This is most important, since owing to the direction of movement there is no axial or sideways displacement between the tool and gear, and if the slots were in line a series of bands of unshaved areas would be left on the gear flanks. The reverse crown effect is applied to the tool so that it is thicker at the ends than in the centre of the teeth. As the gear is passed over the tool the modified sections on the tool are imparted in reverse on to the gear. The action is complete once the gear has moved a sufficient distance to allow the whole of its face width to pass over the common line of centres between the two members.

The taper is only a modified form of crowning and again can be applied by the tool if the reverse taper is applied to the tool and then the underpass technique used.

The crowning can be obtained by a machine equipped with a suitable attachment similar to that shown in Fig. 91. This is usually a sine bar arrangement and the work slide is pivoted to its centre to permit an endwise rocking movement. One end of the slide is supported by a block which moves along an inclined slot in the sine bar. Thus as the feed slide

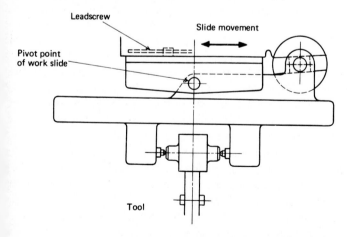

Fig. 91. Typical crowning attachment for a shaving machine

is traversed the end runs up the sine bar which imparts a rocking motion to the slide due to the centre pivot. As the tool traverses across the gear the two are brought closer into engagement thus causing the tool to cut deeper at each end of traverse than at the centre.

The attachment can be used with both the transverse or diagonal techniques but when using the latter it must be appreciated that the higher the traverse angle the smaller the crowning obtained from a given sine bar setting.

A further method of imparting crowning is by the radial infeed technique, but this is seldom used owing to the effect it has on the tool; it involves varying the centre distance between the two members as the feed progresses. It can be seen, however, that this places an undesirable load on the ends of the shaving tool and usually causes the tool to chip and flake at this point.

Restricted or shoulder gears

The pivot point of the tool must pass beyond the extremity of the faces of the gear fully to generate the gear over its entire length. On open type gears having no restrictions the pivot point is usually taken at the centre of the cutter and gear and the stroke then set to allow the points of intersection of the two axes to pass outside the gear face.

When considering restricted gears, however (that is to say gears having large diameter adjacent shoulders which limit the overtravel of the tool), it is necessary to restrict both the crossed axis angle and the pivot point. Fig. 92 shows a typical application and the first step is to determine the position of the interference point A between the diameter of the shoulder and the diameter of the tool. A clearance should be left at the interference point of 1.27 mm and this should always be kept as large as possible since if the tool overtravels at this point severe damage can occur when using the diagonal and transverse techniques. The 90° or underpass method has an advantage in this respect in that owing to the direction of feed's being at right angles to the interference point, the clearance is maintained and never reduced.

Having found the interference point it is now necessary to draw in the line of action and to determine point B, which represents the last point of contact on the end face of the gear — i.e. no action takes place beyond this point. The maximum C.A.A. is now determined by joining points A and B and the width of recess between the gear and the shoulder should be sufficient to allow an absolute minimum of 3° but preferably 5°.

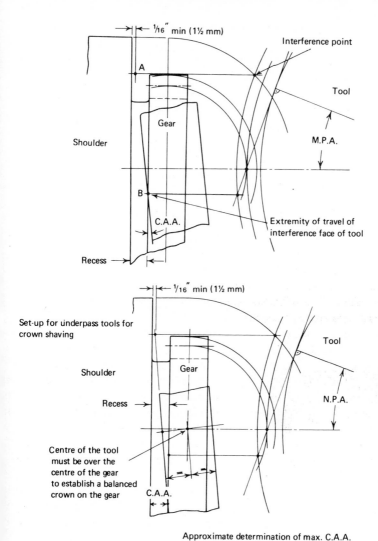

Approximate determination of max. C.A.A.

Fig. 92. Shaving shoulder gears

(e) Preshaving

Although standard types of tools can be used for cutting gears prior to shaving they are not to be recommended. Fig. 93a shows a gear tooth produced by a standard hob and it can be seen that when the shaving tool operates it has to work quite hard with the tips of the teeth to

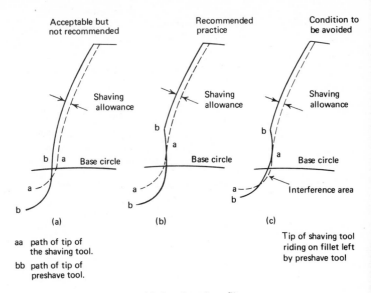

Fig. 93. Pre-shaved profiles

engage the flank of the gear. This causes heavy pressures, flaking of the tips of the tool and a step in the flank of the gear which could be a further stress raiser. Fig. 93b shows a gear cut with a protuberance type tool where the flank is suitably undercut so that the tip of the shaving tool is not required to do any work and a smooth blend is obtained with the flanks after shaving. When producing very low numbers of teeth there is a tendency for the tool to produce a natural undercut in the gear flanks and in this case a protuberance on the tool is not required. Equally, the use of protuberance can lead to difficulty when cutting at high feeds or speeds owing to the low pressure angle on the protuberance itself. When the angle is low there is little side relief and the tool tends to rub at this point causing excessive wear on the protuberance point itself — for this reason some manufacturers prefer not to use this type of tool since it inhibits the production rate under certain circumstances.

Whether protuberance is used on the tips or not the most important rule to be observed is that the shaving tool and the preshaving tool should be designed to suit each other. Fig. 93c shows what can happen if this is not done — a protuberance hob has been used to undercut the flank of the gear for the tool but, since they were not designed in conjunction with each other, the shaving tool contacts the gear on the fillet.

This contact of the tool with the generated fillet can cause the tool to ride the flanks of the gear, at the expense of machine rigidity, and cause errors in the shaved gear.

Correct geometry between the shaving tool, the gear, the pre-shaving tool and the mating gear is essential if the desired results are to be achieved. Theoretically a different protuberance is required for each different number of teeth in the gear to be produced; fortunately a range of teeth can be covered by one pre-shaving tool providing it is designed to suit. The undercut produced by a given protuberance on a 20T gear would occur much lower down the flank on a 100T gear and this in itself may not be objectionable (depending on the fillet condition) providing the shaving tool is made to suit.

(f) Stresses and distortion

The machining of a blank sets up stresses which are normally relieved in heat treatment in the form of distortion. Cold working of a gear blank can occur through taking heavy cuts and using blunt tools and this can take place during the machining of the bore, faces and diameters as well as during the cutting of the teeth. The surface material is therefore highly stressed and the imposed stress is not necessarily uniform around the gear; consequently, when the gear is heat treated, the distortion is not uniform and it is difficult to allow for this unequal distortion.

As the shaving operation is performed between gear cutting and hardening, however, most of this highly stressed surface material is removed, as the shaving operation is relatively free cutting and if correctly performed actually constitutes a light scraping of the flanks of the gear. The amount of material removed is usually approximately 25 — 50 microns per flank and with correctly designed tools the efficiency of the cutting operation is sufficient to remove the highly stressed surface material without cold working the flanks.

It should be realized that the tool works more efficiently on a rough surface as the cutting edges have protrusions on which to bite. A tool will not cut on a highly finished surface, as it can only cut by locally deforming the surface in order that the cutting edges may get below it to remove material. To get this local deformation on a good surface requires heavy pressures and usually results in the gear becoming cold worked. If this is the case, one of the benefits of shaving is lost, since the stressed material is not removed; this results in distortion over and above that inherent in the basic metal itself. Correctly applied, therefore, it could be said that gear shaving helps to reduce the high surface stresses set up in prior machining operations.

The major objection to gear shaving is the probability of distortion in heat treatment subsequent to finishing the gear tooth form. Where gears are not heat treated, or when shaving is the final operation, the problem does not arise and shaving can be introduced without difficulty. Alternatively gear shaving can be carried out after heat treatment providing the hardness does not exceed 40 C Rockwell. The majority of gears are heat treated after the shaving operation and consequently due allowance must be made for heat treatment distortion. One of the first steps, when one shaving tool is used for several different gears, would be to ensure as far as possible that they have the same characteristics of behaviour in heat treatment. Wherever possible, it is advisable to obtain all materials from one supplier and forgings should be of a fine constant grain size. One of the reasons for the successful gear shaving encountered in America is the ability of the manufacturers to stipulate and obtain forgings of a specified grain size.

When introducing gear shaving into the production line it is advisable to consider carefully each gear in the production programme. Most gears lend themselves to the shaving method but the actual section of the gear can be the limiting factor as gears of weak or flimsy section are difficult to heat treat without considerable distortion. The distortion can be allowed for in shaving but the distortion experienced should be as constant as possible and within reasonable limits. Gears integral with short shafts would be more stable than gears having thin unbalanced sections, and it therefore follows that careful selection at the planning stage of gears that are most suitable for shaving will pay dividends.

It is now necessary to issue an experimental batch of gears from a representative batch of the material to be used on the production line. This batch should be as large as economically possible in order to obtain a fair average result. These gears are issued and processed in the usual manner and finish shaved to the required size. Each gear should be carefully numbered and several teeth on each appropriately marked at reasonable intervals. After shaving, these gears should be carefully checked for accuracy of lead, involute and tooth size and wherever possible graphical records taken. If the graphs are appropriately marked, fairly accurate prediction of the distortion can be made by comparison of the graphs taken before and after heat treatment. The usual form taken by the distortion is for the lead to unwind and the involute to grow plus at the tips, and the values are likely to be of the order of 25 microns per 25 mm of face width for the lead and 13 microns to 20 microns plus on the involute.

It is important that the experimental batch of gears put through to determine the characteristics of distortion are representative of the production gears. It follows that quality control at the heat treatment stage will be such as to obtain consistency as far as possible rather than to attempt to eliminate distortion. Achieving consistency is no simple matter since distortion is mainly due to the release of stresses set up in the machined blank and these stresses are not readily controlled. A careful check should be made, therefore, to ensure that the instructions laid down for heat treatment are rigidly followed.

The required dimensions for the gear at the shaving stage are now known and where the allowance for distortion in lead is excessive it becomes advisable to hob the error into the gear. It is always advisable for the shaving stock to be small and regular and the stock would have to be increased if large errors were to be shaved into the gear. If the gears are hobbed out of lead by the required amount the shaving stock will be uniform along each flank and the minimum of material will have to be removed. The hobbing of the gear with this lead error is a simple matter and involves no extra machining or production times; it involves only calculating new differential change gears for the hobbing machine.

Internal gears can be shaved but the process is extremely limited in that the C.A.A. must be kept to a minimum. Unless the teeth of the tool are crowned the contact between tool and gear all takes place at the ends of the teeth and there is no contact at the centre. The face width of the tool is also kept to the minimum to reduce the degree of interference and there is a tooth limitation between the two members similar to that experienced when cutting gears with a shaper cutter.

(g) Materials and heat treatment

Gears manufactured from case hardening steels show to some advantage in shaving since the depth of case can considerably be reduced as compared to gears which have to be ground after hardening. If a high surface hardness and low depth of case are acceptable then the gear can be cyanide hardened after shaving. Similarly, if minimum distortion and reduced dimensional changes are a prerequisite then Nitriding steels can be used with the shaving operation. The long established favourite of the automobile industry is the carburized gear which gives very little distortion in heat treatment and the gears do not have to be refined again later, apart from the removal of nicks and burrs.

Shaving is a precision finishing operation and as such, if the optimum results are to be obtained, the amount of stock to be removed must not be excessive and the preshaving operation must be held to within reasonable limits.

Shaving is capable of bringing about substantial improvements in the quality of the involute profile, the lead or helix angle and the surface finish. Adjacent pitch errors can be improved to a certain extent in that local high spots and errors that occur within a period of contact between a pair of teeth in the tool and gear will be refined. It is exceedingly difficult, however, to obtain any improvement in the concentricity and cumulative pitch errors owing to the nature of the process.

There is no direct kinematic drive and continuity of motion depends entirely on the contact of the successive pairs of teeth. Therefore control can only be exercised over short arcs of engagement. Again, as the cut is obtained by forcing the two members together radially, the tendency is to remove equal amounts from each side of each tooth. Eccentricity involves removing a minimum of metal at the·low point, a maximum at the high point, and half the total errors from one flank only at $90°$ from the zero. As the shaving process is not sufficiently flexible in the degree of control it exercises, it follows that the degree of correction it can impart to this element is limited.

The datum surfaces used during the preshaving operation should also be used for shaving — it is not conducive to accuracy to change the loca-

Fig. 94. Fine pitch shaving tool in operation

tion points. All datum locating points on arbors and fixtures should be held to a maximum tolerance of 5 microns and the bores should be square to the locating faces. Where gears are integral with their shafts they should be shaved from the shaft centres which should be protected, free from nicks and burrs and true for angle.

The calculation of production times for gear shaving is hardly necessary since most gears with the range 0 − 300 mm diameter can be shaved in cycle times varying between thirty and sixty seconds.

(h) Fine pitch shaving

Gears up to 0.25 mod. and finer can be shaved but special tools are required and the operation is best performed on a machine designed specifically for this kind of work. Figs. 94 and 95 show a fine pitch shaving tool and a Sykes VS4A shaver with auto loading respectively. The

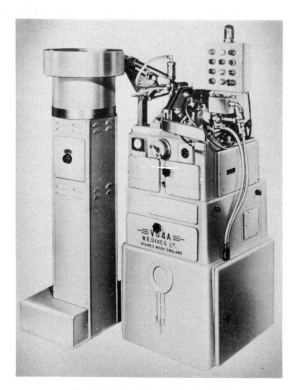

Fig. 95. Fine pitch shaving machine

tool cannot be serrated in the same manner as the coarse pitch tools for obvious reasons and it is usual to grind several grooves along the face width down below the depth of the teeth. The shaving tool then re- sembles a number of thin laminations assembled together and these can be in annular form or as a continuous thread. The disadvantage of the latter is the fact that if the tool is fouled by some obstruction or by an oversized gear, instead of just one tooth chipping or breaking out a whole series of consecutive teeth tend to go. Sharpening of the blunt tool can be carried out in two ways — by (a) regrinding the slits or grooves, and (b) regrinding the tooth profile.

Obviously stock removal on fine pitches is exceedingly small and only limited correction can be carried out. The main advantage is to provide a super finish and to obtain very precise and constant control over the tooth size. On fine pitch gears accurate sizing is important as usually the gears have to run at fixed centres with zero backlash. If a batch of gears are finish hobbed or shaped and are then found to be say 25 microns oversize they cannot be recut and are usually then scrap. The shaver enables these gears to be salvaged as stock removal of this order is fairly easy to accomplish and at the same time a further refine- ment in both accuracy and finish is obtained. Great care must be exercised over the tooth depth to ensure that fillet interference cannot take place and depths are used of $(2.4 \times \text{mod.} + K)$, where K is a value of the order of 25 to 75 microns, depending on the module.

In view of the very high tooth numbers that can result from fine pitches the tools are usually limited to some 75 to 100 mm diameter with 1¼ in bore. They are made in 18% tungsten H.S.S. hardened to 63—64 C Rockwell in the same manner as the coarse pitch tools. How- ever tool life is not so good, even though the stock removal is less, because of the difficulty of providing efficient cutting slots. The softer materials give no problem and good results can be achieved in brass, aluminium and the free machining steels but many fine pitch gears are made in stainless steel and this can give rise to difficulties. Heavy pres- sures are required to remove metal, work hardening takes place very rapidly and tool life generally is poor.

The crossed axes angle is 15° and sometimes higher and the same limitations apply as in coarse pitch shaving with regard to restricted shoulder gears. Only one work slide is used so that only transverse shaving is carried out; the refinements of the other two techniques are not necessary. Again as in coarse pitch work the cut is applied at the end of each pass and a number of cuts usually in the order of two to six can be taken with infeed increments in the order of 5 to 25 microns.

The work holding must receive close attention particularly in view of the high speeds involved. For example, if a 75 mm diameter tool were used to shave a 6 mm diameter gear at 80 metres/min surface speed, the tool would rotate at some 320 rev/min and the gear at some 3700 rev/min. Great care must therefore be taken in the preloading between centres to avoid the centres' burning up during the shaving operation.

1.3.2 Gear lapping

This is a technique developed for the refining of gears which will reduce minute errors in profile, lead and under certain conditions spacing and eccentricity. The obvious advantage is that it will correct these errors *after* heat treatment whether they are due to distortion or whether they were there prior to heat treatment. The action and limitations are very similar to the gear shaving process but, whereas shaving removes metal by scraping or cutting, lapping is pure abrasion and requires a lapping medium to be effective. There are two basic methods used in lapping – (a) running and lapping a mating pair of gears together, and (b) running the gear with a lap. The former is highly favoured for bevel gears of all types, while the second method is used where the gears are required to be interchangeable with each other.

We have already seen from the previous chapters on gear shaving that when two gears are rolled together there is a variation in sliding velocity down the flanks and that pure rolling only takes place at the pitch line. If the axes are crossed, however, a further component of sliding is set up along the axis of the teeth which further improves matters. Lapping can take place therefore either in (a) parallel axes (in which case a reciprocating movement in the direction of the gear axis has to be applied to compensate for the lack of sliding at the pitch line) or in (b) crossed axes (where, although a reciprocating movement is still required, it need not be as fast as that required at parallel axes, but must travel far enough to cover the whole face width of the gear). This last point is best appreciated if the lap is considered as a shaving tool without slots and the metal removal is then obtained with the abrasive compound. The same limitations with regard to direction and length of movement and value of crossed axes angle apply as in gear shaving.

(a) Parallel axes

Lapping can take place with one, two or three laps each of which is in simultaneous engagement with the work and reciprocates along the work axis as it rotates. The object of the multi-laps is to overcome the variation in the sliding velocities down the flanks which obviously causes different rates of metal removal. This is achieved by changing the meshing pressure angle between the laps and the work but maintaining the base pitch. By designing each lap separately the meshing pitch diameters between the work and each lap are different so that the pure rolling, or 'dead spot', which occurs at the pitch line occurs in a different place on the flanks of the gear.

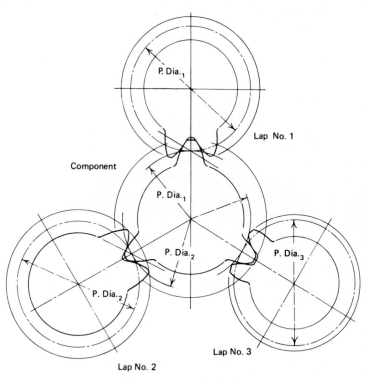

Fig. 96. Gear lapping

This is appreciated by referring to Fig. 96, where it can be seen that there are three meshing pitch diameters and three meshing pressure angles. Thus the positions of poor metal removal between lap 1 and the work and lap 2 and the work are not the same. Lap 2, therefore, tends to remove the metal left at the 'dead spot' by lap 1, lap 3 corrects for lap 2 and so on.

Further balancing of the sliding action down the flanks can be achieved by increasing the rate of sliding or reciprocation along the axis as pointed out in the section on gear shaving (1.3.1); this increases the relative speed of cut at all positions and therefore, although it helps the cut at the pitch line, it also increases the cut above and below the pitch line and does not reduce the differential.

(b) Crossed axes

Lapping usually takes place with a single lap, the axes between gear and lap being crossed as in gear shaving.

Owing to the increased sliding along the axis of the teeth that results from crossing the axes, the rate at which the two members are moved relatively to each other need not be as fast as in parallel axes lapping.

Continuity of contact is maintained by the contact ratio between lap and gear; as there is no direct drive the crossed axes angle must not be too high or control is sacrificed for cutting action.

As in gear shaving the contact point of the pitch line cylinders or point of minimum centre distance must travel across the entire face width of the work before processing is complete.

Gear lapping machines operating on the two lap principle at crossed axes are made by the Michigan Tool Co., and each head can independently be adjusted so that both heads are used to lap one gear or two gears simultaneously.

(c) Laps

These are usually made of cast iron or any other material with a similarly open grained structure which will carry and retain the lapping compound. Laps can be made for both external and internal gears and vary from 25 to 230 mm diameter depending on the application and the machine employed. The lap is kept as large as possible, usually 230 to 250 mm diameter, and when worn is recut with a smaller number of teeth. This may be carried out a number of times.

Lapping is a refining process and it cannot be assumed that it will remove all gear errors. Its principal objectives are (a) to refine the bearing achieved on the gears; (b) to remove slight inaccuracies in profile and lead; (c) to allow the minimum of stock removal; (d) to give improvement in surface finish. Undesirable end bearings can usually be eliminated and with the two lap method high and low bearings on the involute profile can be improved by displacing one lap relative to the other radially.

The time required to lap a gear depends upon the amount of stock to be removed and the type of error to be corrected but on average it lies between thirty seconds to two minutes per side. Long lapping cycles are not desirable and should be avoided since, owing to the variation in cutting efficiency down the flanks, accuracy can be destroyed whilst attempting to improve it. Stock removal can be obtained either (a) by reducing the centre distance between the lap and the gear, or (b) by providing a braking action to the driven member.

The first method processes both flanks simultaneously and since the lap is forced or wedged into the tooth space of the work it has the optimum effect on the reduction of spacing errors and eccentricity. The second method involves some sort of tangential braking to the driven member so that each flank of the work is processed separately and no reduction in centre distance is necessary. Obviously with this method the addendum of the lap must be longer than the addendum of the mating gear, i.e. the lap must engage lower down the involute profile than the tip of the mating gear. Brake pressures may be applied in a number of ways. Probably one of the best is the technique of using a hydraulic gear pump in the work tailstock which is driven by the work as it rotates with the lap. By providing a relief valve the load can be varied as required and is constant for both directions of rotation.

The efficiency of the cut and the results produced depend upon the lapping compound, the pressure of cut and the feed and speed used. Speed of rotation should be relatively low in order to avoid the lapping compound's being thrown clear of the lap by centrifugal action. High speeds of reciprocation along the axis reduce the time required for lapping but the lower speeds give a better surface finish. The lapping compounds vary (they may be oil or water soluble) while the coarseness of the compound may be varied either to speed up the process or to improve the surface finish.

Internal gears may be lapped by the single lap process and where the diameter is so small as to exclude the use of a lap smaller than the gear the lap is made with the same number of teeth as the gear. The same applies to external gears or splines, having stub teeth and high pressure angles; the lap will not roll with the work under these conditions. The lap is therefore made with the same number of teeth as the work and provided with sufficient clearance for the teeth to be easily engaged. As they rotate together the lap is reciprocated along the axis of the work and a braking action applied as the lap drives the work. Internal and external laps can be used in this manner and they are usually held in a floating holder such that the lap can align itself easily to the work. Obviously when the lap and work have the same number of teeth the axes must be parallel, but, if required, the axes can be crossed when lapping internal gears where there is a difference between the number of teeth.

When using two laps it is usual to have one lap power driven and the second driven by the workpiece itself. The time cycle may be varied by means of suitable timers and the whole cycle is fully automatic.

1.3.3 Gear burnishing

This is a cold working process and not a cutting one, the object being to improve the surface finish of the gear rather than to correct errors. Any machine-cut gear has surface irregularities consisting of high spots or peaks which occur either down the involute flanks or along the helix angle. These high spots are depressed in the burnishing process and the gear flank is rolled into one smooth surface. The process does not remove errors from the gear in the accepted sense but of course any surface peak or irregularity is in itself an error so in this sense it can be said there is an improvement in accuracy. The process does not remove material and it is unnecessary therefore to leave material on the gear flanks prior to burnishing.

Three burnishing gears are used each of which is in simultaneous contact with the work. This itself is not located by a work-holding fixture but by the contact of the burnishing gears themselves, in a similar manner to centreless grinding. As described under gear lapping (Fig. 96), the gears are made to operate at different meshing pressure angles with the workpiece and this helps to ensure a better and more even distribution of the burnishing pressures along the flanks of the work.

The arrangement of the burnishing machine is such that the first burnishing gear is power driven, the second one merely acts as an idler and rotates on the stud on which it is held, and the third burnishing gear also rotates on its stud. However this latter stud is in turn mounted on a moveable arm which can be loaded with suitable weights. The pressure for burnishing is therefore obtained through this third gear and the fulcrum arm carrying the weights. The amount of pressure necessary is governed by the diameter and the face width of the gear being burnished and it is extremely difficult to lay down specific rules regarding the amount of pressure required. The following table is merely intended as a rough guide to pressure and duration of the time cycle.

The best method of burnishing is to experiment until the required degree of surface finish is obtained. The smallest pressure should be applied first and the weights gradually increased until the required finish is obtained.

Time and weight required for burnishing

Number of teeth in the gear being burnished (inclusive)	Time of each direction (seconds)	Number of additional weights
13 to 15	5	None
16 to 18	8	None
19 to 21	10	1
22 to 24	12	2
25 to 27	15	3
28 to 30	15	4
31 to 33	20	5

It should be remembered that the time required for burnishing is governed to a certain extent by the pressure applied and the greater the pressure, the shorter this is. Excessive pressure of a long duration should be avoided if possible, since this has the same effect as overloading the gear. The general effect, if the pressure applied and the time cycle are too great, is to throw up an excessive burr at the edges of the teeth. The speed of rotation of the driving gear also plays a large part in the efficiency of the burnishing operation. As a general rule, when using 127 mm diameter burnishing gears, the speed of the driving gear should be approximately 150 rev/min and this speed is fixed and constant on the Sykes machine. The speed of rotation of the burnishing gears should not be too high, since there is a tendency for the oil to be thrown from the contacting surfaces of the gear teeth owing to centrifugal action. Care should be taken to see that sufficient lubricant is applied to the teeth during burnishing as insufficient may result in the gear teeth becoming excessively hot and the flanks burnt.

It is essential that some form of lubricant be used on the gear teeth; this should be a solution consisting of four parts of paraffin and one part of machine oil. This generally reduces the time required for burnishing and should produce a good smooth surface.

The burnishing gears are made in sets of 76 mm high speed steel, hardened and ground to master gear accuracy. They may be reground some four or five times before being discarded.

Burnishing takes place in both the forward and reverse directions and the time cycle can be varied by means of suitable cams which when set give an automatic cycle.

Fig. 97. Gear burnishing machine

Fig. 97 shows a Sykes model gear-burnishing machine complete with burnishing gears, but it should be noted that this is now rapidly becoming an obsolete process and the machine is no longer produced by Sykes. The process is essentially a finishing one and is described since the technique is now being used again in a more sophisticated form for the cold rolling of gears.

1.3.4 Hard honing

Unlike burnishing, hard honing is of fairly recent innovation. A high percentage of scrap occurs in heat treatment of the gears through nicks and burrs on the teeth and the hard honing process enables such gears to be salvaged and indeed even may bring about minor improvements in accuracy.

An abrasive impregnated hone in the form of a helical gear is used which runs in mesh with the work piece at crossed axes. Again the process is very similar to gear shaving — the tool drives the work merely by the continuous contact of the teeth. The conditions of cutting are the same in that the resultant speed of cut is developed from the relative sliding down the flanks due to involute action and the slip along the helix due to the crossing of the axes. The hone and the gear are run together at high speed while at the same time the hone is traversed across the face of the gear in the direction of the work axis.

The process is suitable for external and internal gears both spur and helical, and the limitations are as in gear shaving. It will remove burrs, furnace scale and minor heat treatment distortions, as well as generally improving surface finish. It is an abrasive action and therefore does not cold work the surface of the gear or raise the surface temperature as much as grinding because the metal removal is considerably less.

The two basic kinds of hone available are (a) the plastic cast matrix type (which is impregnated throughout with a silicone carbide abrasive of any specified grit size 60 − 280) and (b) the metal bonded type (which has a similar abrasive bonded onto a solid steel body). The former is dressed on the outside diameter periodically as the hone reduces in tooth thickness. This is done in order to avoid the addendum of the hone becoming so long as to foul the root of the gear being honed.

Fig. 98. Precision Gear hone

The second type is reconditioned by removing the original coating as it becomes worn and renewing the abrasive. This has to be carried out by the hone manufacturer who has the necessary facilities. This type is preferred where the application is such that the hones are liable to breakage and tooth strength is a critical factor. The plastic kind are usually thrown away at the end of their useful life, this usually extending to the processing of some 5000 to 8000 gears depending on the application.

A coolant is used in the process in the form of a honing oil and there is no tendency for the hone to load up. However stock removal should be kept to a minimum such as 25 to 50 microns over pins.

The main advantage of the process is its ability to salvage heat treated gears which have been scrapped owing to nicks and burrs on the teeth, though its ability to reduce errors is very limited. If the gears represent a considerable investment in terms of production time, materials and so on, it is a question of economics as to whether retaining this investment is worth the expenditure of honing machine and hones.

Figs. 98 and 99 show a gear honing machine and hone made by Precision Gear Machines and Tools. The hone is spring loaded into tight mesh with the gear and the pressure can be adjusted; this allows the hone to follow eccentricity in the work or large burrs without breakage.

Fig. 99. Precision Gear model GHD.12 gear honing machine

At the same time, however, it applies enough pressure to reduce these irregularities to an acceptable value. It is essential that the abrasive honing dust be continually flushed away from the work area and a copious flow of coolant is required.

Fully automatic cycles and automatic loading are available and honing cycles vary between ten and 100 seconds each. The procedure for set-up and work holding fixtures is the same as for gear shaving.

1.3.5 Fellows gear finisher

One of the latest methods to be introduced for finishing gears is the Fini Shear of the Fellows Gear Shaper Co., U.S.A. This is a combination of known techniques all put together to give an entirely new recipe. It depends on the following basic principles, all of which have individually been tried and proven in the past but never in this combination:

(a) crossing of the axes of tool and work (of gear shaving);

(b) a simple flat faced gear shaper cutter (cf gear shaper);

(c) use of a helical guide to control the work (cf shaping);

(d) a positive geared index drive between tool and work (cf shaping and hobbing);

The general principle of the process is illustrated clearly in Fig. 100. This shows again the free cutting action which can be obtained by the crossing of the axes. As the tool and gear rotate in the necessary time relationship the gear is moved along its own axis such that the whole face width is processed once it has passed across the cutting edges of the tool. These cutting edges lie in the plane XX and the teeth of the tool engage in a tooth space of the work and sweep across the face of the flanks shearing off the metal in fine chips. The first point of contact at a and the point moves progressively along the line of action through the pitch point b to the last point c.

These points when translated back across the cutting edge of the tool and to the face width of the gear give a path of contact on the gear flank as shown in Fig. 100. A high peripheral speed of the tool is used to give the cutting action and the feed rate is kept low, of the order of 50 to 250 microns per revolution of the work. The crossed axes angle is usually around $15°$ as in gear shaving and the same sort of limitations apply in that values down to $5°$ can be used and shoulder gears can be produced within reason.

The gear must be fed across the cutting edges of the tool at a controlled rate and its direction of movement must be very precise. When

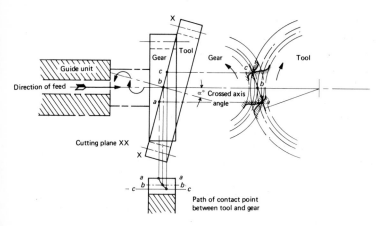

Fig. 100. Fellows Fini Shear method

producing spur gears or components with straight teeth this involves no problem but when producing helical components the gear must be moved along a helical path equivalent to its own developed helix. This is a decided disadvantage in that a different guide is required for each lead to be produced as, although the helix angle of a range of gears could be common, the lead varies with the gear diameter. If two gears of 30° H.A. and the same D.P., P.A. etc., were to be produced, but the respective numbers of teeth were twenty and twenty-one then the leads would be different and the guides would have to be changed. This is different from the gear shaping process which uses a guide to control the cutter, so that, as long as the helix angle is constant, different numbers of teeth in the gear can be produced without changing the guide.

As this is a finishing process the gear has already been roughed out when presented to the tool, so that 'pitching in' has precisely to be controlled. This is achieved by rotating the tool and gear together with the work clamped under low pressure until it positions itself relative to the tool and then it is securely clamped at high pressure. The work is then retracted along its axis to the start position which involves the tool's cutting edges lying clear of the leading edge of the blank. At this point the centre distance is reduced in order to provide a cut and one pass is sufficient to finish the gear.

The process has certain inherent advantages over gear shaving:

(a) It is capable of far heavier removal of stock — 0.250 to 0.600 mm over pins.

(b) The cutting tool is far simpler and cheaper and can be honed in position on the machine or can be removed and face sharpened on a conventional tool and cutter grinder.

(c) Under certain conditions it is possible to use carbide tools which respond favourably to this type of cut.

(d) Gears which are impossible to shave can be produced by this process owing to the fact that it has a controlled index drive and does not rely on the continuous contact of the teeth.

(e) Owing to the positive drive it has more control over the correction of errors, particularly in regard to adjacent and cumulative pitch errors.

In the last two respects particularly this process offers more versatility than shaving and gears having a low contact ratio with the tool are no longer a problem. Unshavable components such as high pressure angle, stub depth splines with a low number of teeth can therefore be produced just as easily owing to the controlled index drive.

It does not necessarily follow that the gears have to be of involute form; components of cycloidal or special form can be produced provided that they are capable of being generated by a circular type of tool. The tool itself is made with a very low top angle of some $3°$ or less; this means that the whole of the form is ground back at this angle to provide side relief when cutting. As the tool is sharpened, therefore, the diameter does reduce, but not at the same rate as its equivalent gear shaper cutter which uses angles in the order of $6°$ to $10°$. If the gear to be produced is of non involute form this is particularly important since the change in diameter of the cutter from front to back must be kept to a minimum. When it is stated that non-involute forms can be produced, it must be appreciated that there are still limitations but these are not as great as by conventional methods; under these circumstances the top angle is sometimes reduced to zero.

The advantage of the involute lies in the fact that as the diameter of the tool changes the basic rack required to produce the component is still conjugate to the tool. This is not so with non-involute forms and the basic rack varies with the cutter diameter. Theoretically the cutter should have a different form front to back to compensate for the change in diameter when providing non-involute profiles. For this reason the top or form angle is kept as small as possible in order to keep to a minimum the change in form produced on the component. A very precise control of the kinematics is required on the machine in order to give optimum results and this involves accurate ground gears in the index drive

This precise control of the index drive enables the tool however to improve the pitch characteristics of the component since unlike shaving the process is no longer dependent on the number of teeth in engagement at any one instant.

The pitch characteristics are now very similar to those produced in gear shaping since the same basic elements are involved:

(a) accuracy of the index of the work;

(b) accuracy of the index of the cutter;

(c) accuracy of the cutter tooth spacing.

It follows, therefore, that the degree of accuracy obtainable should be comparable to that produced on a high grade shaper when finish cutting. The average time cycle is similar to that of gear shaving and is neither faster nor slower, the time varying between fifteen and sixty seconds depending on the component. The degree of accuracy claimed by the Fellows company is as follows:

Profile 10 microns, lead 5 microns, 13 microns pitch line run-out, adjacent pitch 10 microns — size within 13 microns

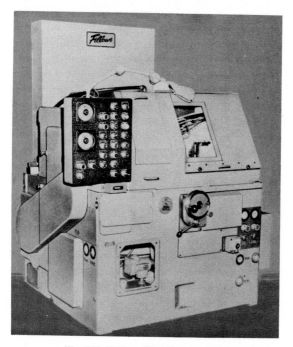

Fig. 101. Fellows Fini Shear machine.

Obviously these are average figures and can be better or worse depending on the component.

Owing to the extremely rapid production cycle the machine is of necessity equipped with quick change work holding fixtures or with automatic loading. If necessary minor adjustments can be made to the lead produced by the guide unit in order to allow for changes due to heat treatment distortion and so on, while crowning can be achieved if required.

Fig. 102. Typical tool and component

Fig. 101 shows the Fellows machine and Fig. 102 a typical tool and non-involute type component. Fig. 103 illustrates the accuracy both before and after the finishing process for a typical component which has been only half processed through its face width. This illustrates quite clearly the nature of the cut and the degree of surface finish obtainable.

1.3.6 Gear grinding

Grinding is essentially a finishing operation (although fine pitch gears are sometimes produced from the solid) and has the advantage (a) of

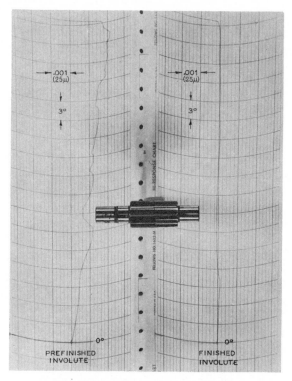

Fig. 103. Graphs of accuracy produced

being able to work hard materials and (b) of giving a finish of very high quality and accuracy. There are a number of different processes available and these can roughly be classified into the following basic groups:

(a) form grinding;

(b) generating with single or twin wheels;

(c) generating with a threaded type wheel.

(a) Form grinding

This type of grinding usually utilizes a disc wheel which grinds both flanks of a tooth space at the same time and is similar to milling in that it does not generate the form but requires it accurately to be produced on the tool or wheel before processing starts. The process is best suited to the production of spur gears although helical gears can be produced on certain machines under suitable conditions. The wheel can obviously be made to grind the flanks only and leave the root of the teeth untouched or it can grind the root at the same time as the flanks. If twin grinding wheels are used they may be arranged so that each grinds one flank of a tooth space only since this reduces the load on each wheel.

The involute profile varies with each number of teeth to be produced and is usually dressed on to the wheel by means of a pantograph type dresser. Master templates are first manufactured and placed on the dresser, the arms of which contain suitable diamonds which operate through the linkage to produce the dressed form on the grinding wheel. It is essential that the wheel be set on the centre line of the gear and that the template be accurately aligned to dress the wheel symmetrically about the centre line of the space. Fig. 104 shows the methods of form grinding external spur gears: (a) single wheel clear of the root; (b) single wheel grinding root and flanks; (c) twin wheels grinding the flanks, and (d) a typical twin arm pantograph.

Most gears require more than one wheel dressing whilst being form ground and the number depends on the material and its hardness, the number of teeth and face width of the gear, diameter and grade of wheel, and the amount of stock to be removed. Stock removal over the

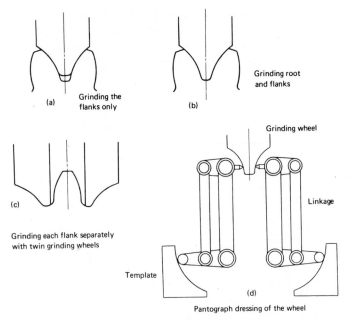

Fig. 104. Methods of form grinding

amount necessary to remove the errors in the gear from the rough machining operation and heat treatment distortion should be avoided, since it increases the production time, reduces the depth of case and in extreme circumstances can temper the hardened zones locally. The modern grinder usually has some automatic means of applying the cut and each space can be roughed out to a size before indexing or each cut increment applied after each work revolution, which brings all teeth to size at nearly the same time.

The grinding stock should be in the region of 130 microns to 250 microns for gears up to 250 mm diameter 4 mod., and the amount per cut should be 10 microns to 18 microns for finishing and 18 microns to 30 microns for roughing the same size gears. Grinding of internal gears is possible with a special attachment, as shown in Fig. 105, but obviously the internal diameter cannot be too small since both the wheel and the drive mechanism lie inside the gear. Owing to the very large area of the wheel in contact with the gear this type of grinding is prone to over-. heating and care is required to avoid 'grinding abuse'.

Fig. 105. Grinding internal gears

(b) Generating

This method is superior to the form grinding process in that only a limited portion of the wheel is in contact with the work at any one time, thus reducing the tendency to overheat and making dry grinding possible. The most common of the generating type machines are (a) the single conical wheel type; (b) the twin wheel type set at generating P.A.; (c) the twin wheel set at zero P.A.

The first type uses a single reciprocating wheel which moves back and forth across the face of the gear as it slowly rotates. Several strokes of the wheel are made per incremental roll depending on the feed rate used, producing a series of flats on the involute profile. However, as in the other generating processes already described, these flats are so close together as to appear a continuous curve. The method of obtaining the rolling motion of the work varies and can be by a master rack and pinion or by a drum and tape mechanism. Whichever motion is used the master pinion or drum is equal to the generating pitch diameter and the grinding wheel is dressed to the equivalent generating pressure angle, as shown in Fig. 106.

Thus the grinding wheel head is set to reciprocate back and forth through the face width of the gear when the axis of the generating drum is on the line XX. At this point also the tips of the flanks of the tooth space to be ground are at positions a and b respectively and the band is in contact with the drum at c. When the feed or generating motion is started the point c moves progressively across to c_1 and c_2 where generating ceases. The points a and b on the tooth flanks now occupy the positions a_1 and b_1, and the complete tooth space has been generated. Now indexing may take place or the slide may be returned to its start position.

Internal gears cannot be ground by this method but it is suitable for helical gears, the wheel head assembly being swung to the helix angle so that the wheel reciprocates along the helical flanks.

The dressing of the wheel is much simpler than in the form grinding process since it is necessary only to dress straight sided at the pressure angle and the number of teeth in the gear has no effect on the angle.

The generating grinders manufactured by the Maag Company in Switzerland are probably the best examples of the twin wheel type both in the zero and generating pressure angle methods. In both cases the wheels are made concave so that only a narrow rim on the wheel is used to give theoretically point contact. Once a cut is applied, of course, this

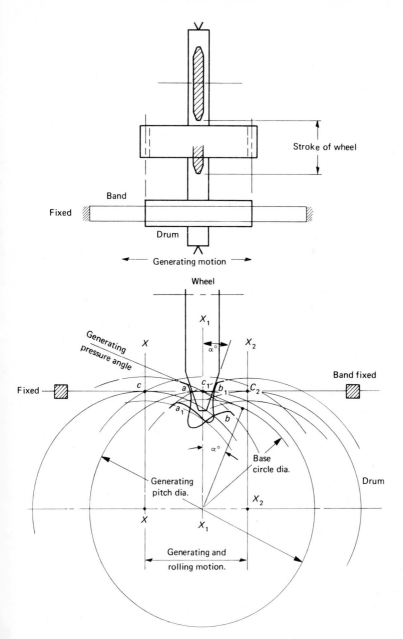

Fig. 106. Generating type gear grinder

theoretical point becomes an elliptical contact area depending on the depth of cut. Fig. 107 shows the general principles of the two methods, and it can be seen that with the generating P.A. principle the wheel represents the flanks of a rack where the angle is equal to the P.A. at the generating pitch diameter. The zero pressure angle method is similar in that since the pressure angle is zero the generating pitch circle diameter is equal to the base circle diameter.

If it is necessary to grind the root of the gear and possibly to form the fillet in the root, then the generating pressure angle method must be used since the zero angle can be used only for grinding the profile down to the base circle diameter. Any blending required below this diameter must be done by the gear cutting tool at the machining stage prior to grinding.

Fig. 107 shows the fundamental difference between the generating and zero pressure angle methods, the former being shown in (a), (b) and (c) and the latter in (d), (e) and (f). In Fig. 107a the twin wheels set at the P.A. α° can be clearly seen. The wheels are dressed to give a small land only on the tips so that just a thin area on the rim is used. The first contact point between the wheel and gear flank is at a and in Fig. 107b this contact point is reproduced on the face width of the gear by the radial line $a_1 a_1$ of the wheel.

As the gear rolls relative to the wheel, however, the contact point moves to c which corresponds to the radial line $c_1 c_1$ on the wheel, and this contacts the face width on both the leading and trailing arcs of contact. The contact point c on the involute is therefore reproduced by the wheel at two positions on the face width. This involute contact point between the wheel and the gear moves progressively up both the gear profile and the flank of the wheel so that contact starts at a and finishes at b. Apart from the contact point a all the successive points have two positions of contact on the face width which vary in their axial positions as well as down the profile. The wheel is therefore cutting in both the leading and trailing arcs of contact and the effect of this when coupled with axial feed is to give the diagonal or criss-cross pattern peculiar to this method. This is clearly demonstrated in Fig. 107c.

The conditions are entirely different with the zero P.A. technique, in which the work is all done by the very tip of the wheel, theoretically at one point. Fig. 107d shows the successive contact points on the involute $a - b - c$ but it should be noted that the contact point on the wheel does not change. This has the effect in Fig. 107e of the wheel cutting at one point only and it does not have a leading or trailing arc of contact.

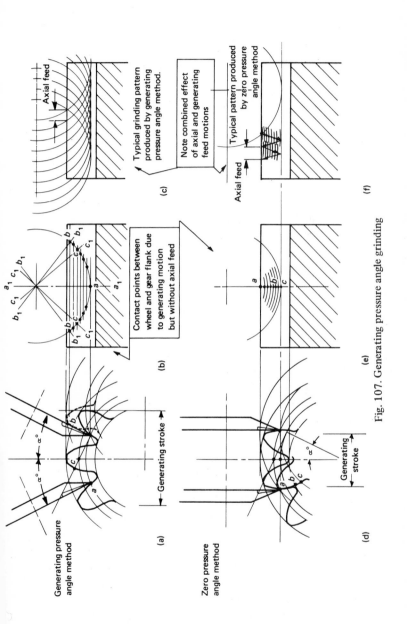

Fig. 107. Generating pressure angle grinding

When the rolling motion is combined with axial feed we have the effect shown in Fig. 107f, the angle of the diagonal grinding pattern varying with the amplitude of the axial feed. In practice the theoretical point contact is slightly extended so that there are short arcs of contact along the face width giving the illusion of a wave along the flanks.

The zero method has certain advantages in that faster axial feed rates can be used and profile modifications in both planes can be effected. Modifications to the profile in either the involute or helical sections cannot be carried out with the generating pressure angle technique owing to the changing point of contact on the wheel. Fig. 108 shows a helical gear being ground at a generating P.A. $\alpha°$ and helix angle $\beta°$ and it can be seen that the straight line generators proceed diagonally across the face width of the gear. If we consider three generating rolling positions 1, 2 and 3, we must relate these positions to the face width at section AA and to the equivalent point A_1 on the generating helix section XX.

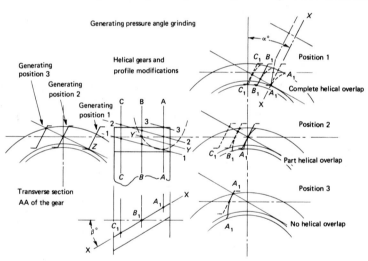

Fig. 108. Profile modifications

Due to the helical overlap action the straight line generator XX contacts the involutes at sections BB and CC whose equivalent involute positions relative to the generator are shown as B_1 C_1. It can be seen that the contact position has changed.

If the grinding wheel were of infinite diameter, therefore, it would contact the helical gear at all positions along the generating line 1 1; in other words the contact point on the involute and along the helix is changing. By the time the gear has rolled to generation position 2 it still has part helical overlap in that it is still in contact at section BB but ceases contact at a section part way between B and C.

When it reaches generation position 3 there is no helical overlap and the contact only occurs at section AA and prior to it.

All conditions are met therefore while the straight line generator XX is maintained, but if local modifications to profile are required this necessitates deviations from the plane XX.

If it were required to provide a 25 microns minus modification at point 2 on the gear flank it would be necessary to make the tip of the wheel plus 25 microns at this point. Since for obvious reasons the grinding wheel must have a diameter then the point 2 takes the path shown by the dotted line on the diagram.

The point Z therefore contacts the pitch line at section BB when the gear has rolled to the generating position Y. Since the point Z is plus 25 microns relative to the straight line generator XX it follows that in addition to removing 25 microns at Z section AA it also removes metal at the pitch line on section BB which is of course unacceptable.

It can be seen, therefore, that profile modifications cannot be produced by this method owing to the changing point of contact between the wheel and gear flank. This does not happen with the zero generating pressure angle method since the contact point is always at the same place on the wheel periphery and modifications can be effected by changing the position of the wheel relative to the work; Maag carry this out very effectively, displacing the wheel along its own axis. The idea is to locate exactly the contact point on the flank at each co-ordinate of the axial travel and generating stroke — this is necessary in order to synchronize the displacement of the wheel. Each position of the generating stroke corresponds to a point of contact and serves as the prime mover for the profile correction — whilst the axial feed motion acts as the prime mover for correction along the length of the tooth.

A transmitter sends an impulse through a hydromechanical copying system to a receiver which scales the impulse to the correct degree.

Fig. 109 shows the basic kinematics of a Maag type grinder, the axial feed, generating and indexing motions being located at the front of the machine.

The column can be swung to the helix angle at the generating pitch diameter and contains a cross beam which can be adjusted vertically to the gear diameter and which carries the two grinding heads. The gear to

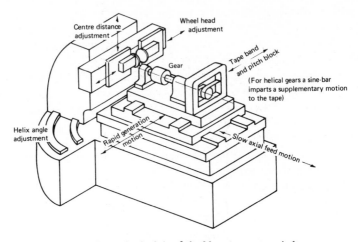

Fig. 109. General principle of the Maag type gear grinder

be ground is mounted horizontally between centres and the pitch block is integral with the arbor carrying the gear. The pitch block is rolled on the tapes causing the motion necessary to generate, the length of stroke and speed being dictated by the crank disc. The feed carriage, carrying the generating slide, slowly reciprocates the gear past the wheels to effect the generating of the face width. When one pair of flanks has been ground the motions cease with the wheel clear of the work and indexing takes place from a master divide plate. When grinding helical gears a further supplementary motion is imparted by means of a sine bar mechanism, which displaces the tape band in proportion to the axial stroke.

(c) Threaded wheel grinding

This is a continuous indexing technique using the fundamental principle of gear tooth development from a rack. It is analagous to the gear hobbing system in that the grinding wheel is formed with a continuous helical thread in the same manner as a worm and the work is rotated at a set rate in direct relationship to the wheel. We have already seen from the gear hobbing process that certain parts of the rack do more work than others when generating a gear and for this reason the hob is shifted along its own axis in order to distribute the wear more evenly over the hob teeth.

It follows that if the grinding wheel were left in one position relative to the gear being ground, certain parts of the flank of the wheel would wear more quickly than others. It is desirable, therefore, to supplement the generating motion by introducing a further rolling movement of the work under the wheel in order that the full face width of the wheel be explored and the wear distributed more easily. This can be appreciated from Fig. 110 which shows the relative motions and directions of travel of the various principal parts of the mechanism.

Spur and helical gears can be ground by this method and, if required, gears of 1½ mod. and finer may be ground from the solid without prior cutting or forming of the teeth. The same wheel may be used to grind

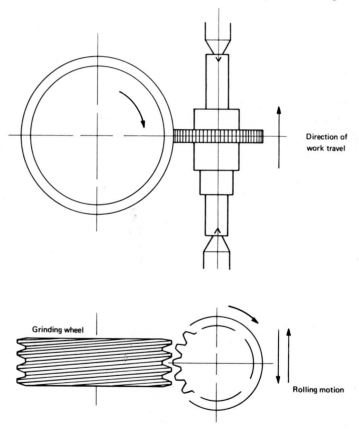

Fig. 110. Gear grinding with threaded wheel

spur or helical gears provided they have the same normal base pitch. Views on methods of forming grinding wheels vary slightly with different manufacturers; some recommend finishing by crushing, others finishing by diamond dressing. The wheel is always roughed out by means of a suitable crusher and is then dressed with a diamond. Some manufacturers use the diamond as a means of finish dressing, others as a secondary stage following up with a final crushing operation to finish the wheel.

Crushing a new wheel could take some four hours and for this reason it is desirable to stock preformed wheels of standard modules and P.A.s so that they need be finished only to individual requirements.

The work is mounted vertically between dead centres and the axis of work and wheel are inclined to suit both the helix of the wheel form and the helix of the gear to be ground. Change gears are included to control the relative motions of lead, rate of feed and indexing and the work is traversed relative to the wheel in order to generate the full face width. The principal advantage of this type of machine is the very good productivity coupled with accuracy, although a finer degree of operator skill is necessary. Feeds and speeds are again very dependent on the general conditions, accuracy required, type of wheel, hardness of job and so on.

(d) General comments

When wet grinding great care must be exercised to clarify the liquid so that dirt and metal particles are filtered out before recirculating. The volume should be adequate to prevent excessive local rises in temperature and the emphasis should be on volume and not pressure. Care is necessary in choosing the correct structure, grit, hardness and bonding of the grinding wheel. Wheel speed and glazing of the wheel require attention, the correct grinding speed and frequency of dressing being of prime importance. The most common complaint is 'grinding abuse' or cracking of the surfaces which can be due to a variety of reasons. The grinding process generates considerable heat at the point of metal removal and if this is not adequately controlled then the surface is tempered and minute cracks develop which can lead to subsequent failure on high duty gears. A 100% check should be made for cracks after grinding and this is usually done by the Magnaflux technique. The etching technique is used to detect signs of surface tempering which usually show up as black areas, whereas areas with no tempering usually show grey.

The advantages of the grinding processes are fairly obvious – (a) they are final operations (so that the accuracy produced is not reduced by other operations), and (b) hardened parts can be processed. The disadvantages are that, compared with other machining processes, they are relatively slow and suffer from operator abuse which is difficult to detect without stringent and expensive inspection.

The grit is the main element of the wheel and this is graded in various standard sizes. The two chief types of abrasive media are aluminium oxide or silicon carbide based (Carborundum).

The mesh of a calibrated screen determines the size of grain and the number is the number of openings per linear inch according to an international scale. The sharp corners of the grit when embedded in the bonding of the wheel give it the required cutting action. The hardness of a wheel is a measure of the resistance of the bond to the grit's being torn out and depends on the strength of the bond and proportion of abrasive. This proportion of abrasive to bonding medium and total volume determine the density of the wheel and its structure; it has a considerable effect on the chips produced and the heat generated.

A method of assessing the suitability of a wheel is to inspect the swarf produced; if this is free, in long curls and not discoloured, it means good grinding. If the swarf is welded together – usually into small balls – and is heavily discoloured, this is indicative of severe local heating possibly from the use of too hard or too dense a wheel.

1.3.7 Chamfering and deburring

When metal is removed at high speed, burrs and rags are produced where the tool breaks out on the edges of the teeth. These burrs are best removed as a separate operation on a machine specially designed and built for that purpose. This is because the degree of accuracy required is low and it would not be economic to tie down a high precision gear cutting machine with attachments to perform such an operation. The deburring machine can also chamfer the leading edge of axial-sliding gears of the type used in automobile gearboxes and lathe headstocks.

Prior to the advent of the special machine tool these burrs were removed by hand either with a file or with special wire brushes. Obviously this was extremely slow and expensive, even though only low grade labour was used. Burrs can be removed very easily by the technique known as 'tumbling' which is extremely fast and economical; unfortunately this is not ideal for the larger type of burr and rags found on gear teeth.

A number of techniques are available for machining the burrs and these include the facility slightly to radius or chamfer the edges to prevent eventual flaking. The following is a rough summary of the various methods available for both the removal of burrs and production of chamfers:

(a) grinding with formed wheel;

(b) cold rolling the chamfer plus set of trimming cutters;

(c) milling – variety of pencil cutters and end mills;

(d) rotating single point tool or tools;

(e) reciprocating single point tool or tools.

(a) Grinding

Different grinding wheels are required for each module and P.A. and they are specially formed to suit the individual gear. In some cases a profiled wheel is used which is plunged into the gear tooth space at an angle so that a uniform chamfer is produced along the edges of the flanks. The wheel is then retracted and the gear indexed to the next space and the cycle repeated until all the teeth are formed. Other processes use a wheel which (although it must be changed for module and P.A.) is suitable for all numbers of teeth and is continuously indexed. This is similar to the hobbing process except that the cut is continuous and not interrupted and the driven grinding wheel causes the gear to rotate around its own axis. The wheel is fairly large to avoid rapid wear – usually some 200 mm diameter, rotating at 3 000 rev/min whilst the time cycle is extremely fast – eight to ten seconds per side.

(b) Cold rolling

An excellent method of removing the burrs from the end faces is by cold rolling with a suitable set of gears. The method is extremely rapid and simple and the tools relatively cheap and long lasting. The work piece is mounted with its axis vertical on a free running spindle and is engaged by two further vertical spindles which are power driven and mounted on a separate head. One spindle carries the deburring cutters which are arranged one above the other and with a gap between them slightly greater than the face width to be processed. The three gear rollers are mounted on the other spindle (Fig. 111), the centre member being helical (when producing helical gears) and the remaining two

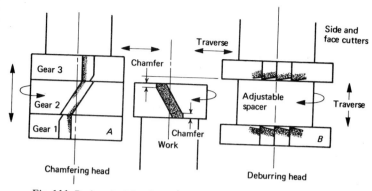

Fig. 111. Basic principle of gear chamfering and deburring machines

being spur. The head carrying the spindles can advance to and withdraw from the work, and the spindles can traverse up and down in the direction of their axes. Once the workpiece has been loaded on to the fixture the head advances with the spindles, rotating until the centre rolling gear engages the work. Once full depth is reached the vertical spindles feed upwards so that the spur roller A engages with the lower face of the gear and forms a small chamfer. At the same time the lower cutter B

Fig. 112. Precision Gear chamfer and deburr machine

on the opposite powered spindle engages the face of the gear and trims off any excess metal pushed out by the rolling operation. The sequence is reversed when the spindles feed down so that the upper faces are deburred and chamfered. The pressures and travel are all adjustable so that the degree of chamfer can be controlled. Fig. 112 shows a machine built for this type of operation by Precision Gear and equipped with automatic loading.

(c) Milling

This is probably the most popular technique and is certainly the most flexible and universal of all the methods. The cutters vary considerably in shape and size from end mills to bell type cutters which cut on their inner surfaces. The pencil type cutters are used mainly on the copying

Fig. 113. Detail of internal tooth chamfering on a tooth chamfering machine

machines where the tool rotates at high speed and is moved in and out axially by means of a cam in order to impart the required profile on the work. Fig. 113 shows a typical example of producing a chamfer on the internal teeth of a clutch on the Hey machine. This type of machine is very flexible in use and is capable of tooth rounding and single or double chamfering on both external and internal gears.

Larger chamfering operations can be performed on the machine by means of a special angular milling spindle. With this method much larger cutters of heavier duty can be used.

Fig. 114. Hey machine. Radiusing teeth with bell type cutter

The bell type cutter shown in Fig. 114 is used for radiusing the teeth of the larger external gears and works on its inner edges to produce two teeth simultaneously. This double working of the two opposed edges saves further set-up time, but care is required at the setting stage in order to avoid the cutter's fouling the gear teeth on the entering and leaving of the cutting tool. It means that one flank is produced by down cutting and the other by up cutting, but the finish produced is acceptable. The indexing is intermittent and takes place when the cutting tool is withdrawn axially and some modification to the cutter profile is usually required in order to produce the required form on the gear.

Twin spindle end mill techniques are used in the mass production of the double chamfer on the clutch teeth of synchro-mesh gears as shown in Fig. 115. The use of twin spindles enables the two chamfers to be produced at the same setting and again this is an intermittent indexing technique; the indexing takes place when the cutters are withdrawn axially. It is essential for the angle of the tools relative to the work to be adjustable, and with this technique in particular some adjustment of the centre distance between the two end mills is essential unless the machine is to be considered as special purpose and tooled for one component only.

Fig. 115. Hey machine – using twin spindle end mill techniques

(d) Single point tools

Here is a very simple means of providing a slight chamfer on the edges of the gear teeth which is suitable for short or long production runs and involves only very cheap tooling. One or two cutter spindles can be used, the latter number being preferred wherever possible as it enables both faces of the gear to be processed simultaneously. The tool is similar to a tool bit on a boring bar but the profile is shaped to sui

Fig. 116. Chamfering with single point tools

he tooth flank of the gear to be produced. The two spindles are independently adjustable and the work must be prelocated relative to the tools in order to provide the chamfer in the correct position.

The tools and the work are rotated in synchronism, the tool revolving once for each tooth on the gear and the gear being completed after one revolution. When larger chamfers are required it may be necessary to take two or three cuts in which case two or three revolutions of the work are necessary. The cycle is automatic and the indexing is continuous. The process can be likened to fly cutting of gears (Fig. 116).

e) Reciprocating tools

A refinement of the previous process is the use of reciprocating or oscillating type tools which are particularly useful when machining gears close to shoulders. Fig. 117 shows clearly the application when deburring and chamfering both faces of the smaller gear of a double gear where only a limited space is available between the two gears. The tools reciprocate and are used in a woodpecker action to machine the lower faces, which are obstructed and cannot be machined in the conventional manner.

A further use of this type of tool is illustrated by the Hey Company in their machine for providing the chamfer on the teeth of starter ring gears. Two versions are available, the first using a single oscillating tool

Fig. 117. Close up of oscillating cutter producing chamfer under
limiting conditions

and the second, with fully automatic loading, using a multiple tool se
up. The latter is capable of chamfering 360 teeth per minute or pr
ducing 180 rings per hour and the tooling is preset so that rapid chang
overs can be achieved.

Fig. 118. Hey machine for chamfering starter rings

PART 1.4: COLD ROLLING

This process was introduced a few years ago and the possibilities it offers over conventional machining are enormous — unfortunately there are also inherent limitations which have not yet been overcome. It is different in one very important aspect in that it does not remove metal in the form of swarf but displaces it under heavy pressures. The process is one of controlled metal displacement by extrusion and as such is more limited than conventional machining methods. There are now a number of machines on the market for cold rolling a variety of forms, all using different techniques. However they can roughly be separated into two types:

(a) those using racks or rotary dies whose axes lie in parallel planes (most suitable for high pressure angle work); and

(b) those using rotary or worm type dies where the axes are inclined at around 90° (capable of producing much lower pressure angles and larger pitches).

Of these the Michigan and Landis machines exemplify the first group in that one uses a rack die and the other a rotary die but the axes of die and work are parallel. (Michigan also manufacture rotary dies for finishing only.)

The second group is covered by the Maag and Grob machines which use rotary and worm dies and different configurations, but which are ideal for producing work in bar or long lengths. They all have one basic principle in common however — the die always operates in the tooth space of the component and the metal displaced in the root by the tip of the die is forced to flow to form the tips of the tooth. The process aims at reducing the metal to a plastic state locally so that it will flow easily, and for this reason it is essential that the material be ductile.

Fig. 119 illustrates the basic idea of the process in that the metal in the area A is forced to flow in the direction of the arrows to fill the area B such that the initial blank diameter D_1 is increased to D_2.

The process therefore is simple in its approach and lends itself to extremely rapid production cycles. Unfortunately, the problem is obtaining the necessary state of plastic flow in the metal without undesirable side effects.

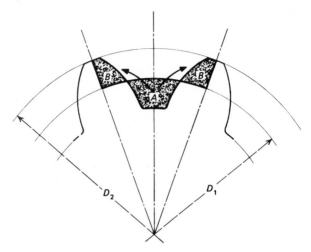

Fig. 119. Cold rolling

1.4.1 Materials for rolling

The success of the rolling technique depends on the physical shape of the part and the nature of the material from which it is produced. In the rolling process the material is stressed beyond its yield point in order to cause it to flow plastically. This is a cold working process and therefore the phenomenon known as 'work hardening' can occur depending on the material analysis, its initial hardness, the number of work cycles and the degree of plastic deformation. Most of the work hardening takes place in the root of the teeth since this is the area of maximum flow and plastic deformation. It is usually superficial and decreases below the surface of the material so that at some 1 mm from the surface the original core hardness is maintained.

Excessive work hardening has to be avoided since it reduces die life, on the other hand it could mean that in certain cases the component could be used without requiring subsequent hardening or toughening operations.

Forms may be rolled in any material which is sufficiently plastic to flow and to withstand the stresses of cold working without disintegration. It should be ductile and homogeneous while the yield strength should not impose excessive pressure on the dies.

Materials with an elongation figure of 12% or more are desirable as a measure of ductility for rolling; however, other features such as hardness and yield strength must be taken into account.

To obtain a permanent state of deformation of the material the yield point must be exceeded; the higher this point the greater the pressure required and the less the tendency to flow.

Although in general the hardness and elongation factor give reliable guides to rollability, care should be taken with certain of the alloys. Some can be soft, show low yield and high elongation yet when rolling may develop high work hardening tendencies which raise the yield point and make cold rolling virtually impossible.

Many of the so called 'free machining' steels are not ideal for rolling since they usually include sulphur, and if this is present in a content higher than 0.13% it is detrimental to surface finish and may cause flaking. Sulphur lowers the ability of the material to withstand cold working. Leaded steels being soft and malleable would appear to be ideal for rolling, but the direct opposite is true since the lead has the same effect as high sulphur content.

The following steels are considered good for rolling:-

AISI 1300 series manganese

3100 series nickel chrome

41000-4100-4300 series molybdenum

8600 series nickel chrome molybdenum

Excessive hardness of the material results in poor die life and generally the hardness should not exceed 28 Rockwell. Where the form to be produced or the nature of the job is considered difficult it may be necessary to reduce this value. As a result of the cold working experienced in rolling, the hardness of the material is increased by some eight points Rockwell and a surface finish of some 10 to 15 microns normally obtained.

1.4.2 Michigan rack process

One of the earliest machines introduced commercially, this kind uses rack type tools; the general idea is shown schematically in Fig. 120, and the Michigan Roto-Flo machine in Fig. 121.

The two rolling racks are tapered so that the initial depth of contact is small and the depth of penetration increases as the racks are traversed and the metal starts to flow.

The racks are mounted on extremely rigid slides travelling in opposite directions whose speeds are synchronized by a pair of master racks and pinions. All the work is done in a single stroke in the one

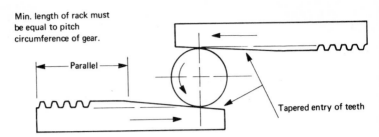

Fig. 120. Cold rolling by means of rack dies

direction and takes some three to five seconds; for this reason most machines are automated to ensure good machine utilization time.

The forming racks are usually made in one of the grades of high speed steel, are all of selected quality and are specially heat treated to a fine grain structure with good impact values.

A low viscosity oil is used to prevent welding of the tool and component flanks and to keep the tool clean. Obviously it must have sufficient additives of an E.P. variety to prevent galling. This method applies high beam loads to the teeth and consequently is only suitable for high pressure angles (i.e., 25° or more) and for stub depth teeth. For this reason it is seen in operation in most plants on components which require the S.A.E. type involute spline form of 30° pressure

Fig. 121. Michigan Roto-Flo machine

angle, 50% stub depth or similar characteristics. It is also ideal for threads or oil grooves or for splines and serrations requiring foolproof assembly characteristics such that one tooth is blocked or removed at certain intervals.

Spur and helical gears can be produced but the same limitations apply as regards depth and pressure angle, and the rack dies are made with the appropriate helix angle. The maximum face width that may be produced is limited by the rack tools, since these have to be the same width as the component, and also by the pressure required, which obviously increases with face width.

1.4.3 Landis circular die process

This is similar in principle to the Michigan except that the forming dies are cylindrical and not in rack form. Two opposed dies are used and the component is placed between them so that the entire face width can be produced simultaneously. It has an advantage over the former process in that the diameter to be rolled is not limited by rack length, and the face width is not a limitation.

These machines can be equipped with 'thru-roll' facilities which enable parts to be rolled in bar form of any practical length. Where the 'plunge' or 'thru-feed' technique is used the dies rotate and drive the workpiece while pressure is applied. Formation of the teeth is accomplished by rapid incremental penetration of the die teeth into the work and with infeed or plunge rolling this is achieved by moving one of the dies in radially towards the other. When thru-feed rolling the work is forced axially through the dies, so that face width or length is not a limiting factor.

It is considered impracticable to roll gears with less than 20° pressure angle measured normal to the helix; the higher angles are preferred since the rolling pressures are lower and the material flow pattern is better, all helping to give better die life. The higher pressure angles give greater beam strength since the teeth are thicker at the base and this gives greater resistance to fatigue and failure. The lower angles result in larger tip lands on the gear and dies which tend to resist penetration, resulting in higher rolling pressures. The rolling pressure and material flow are helped by providing the largest possible radii to the tips and roots of the teeth.

When considering involute forms for rolling, certain fundamental aspects of involute geometry must be borne in mind. For instance, the undercut condition for gear teeth starts at approximately eighteen teeth for 20° P.A. and undercut profiles are not recommended for roll forming.

It is well known that if two gears are rolled together when the profiles are undercut there is a tendency for locking at certain phases of engagement. It is unlikely that the die or component would stand the stress set up at this point, and secondly the material displacement from the undercut would form a bulge on the involute profile above the base circle and adjacent to the undercut. Fortunately the undercut condition is undesirable from the gear designer's point of view, for it shortens the line of action, destroys active involute profile and weakens the teeth. Therefore, steps should be taken to correct the gear so that undercutting does not exist. The undercut deliberately introduced for gear shaving is not necessary during forming by rolling.

The thru-feed method is particularly advantageous when rolling helical gears or parts in bar form where the diameter of the part to be rolled is larger than any other diameter on the workpiece. The pitch determines the rate at which parts can be rolled, but on finer pitches rates up to 2.5 metres/min can be achieved.

Michigan

This is the same as the Landis plunge rolling technique but the

Fig. 122. Michigan cold rolling machine for gear finishing

machine shown in Fig. 122 is intended for finish rolling only from
blanks which have already been pre-cut. This technique has already
made considerable progress in the U.S.A. and is replacing shaving in the
production of automatic transmissions. The two dies complete with
work piece and automatic loading device can be seen in Fig. 123 while
the graphs in Fig. 124 show profile and lead charts which are typical of
the quality which can be achieved by this method.

1.4.4 Maag process

This process uses twin cylindrical dies with their axes inclined at
nearly 90° (depending on the application) to the axis of the component
and is ideal for reproducing components in bar form.

The general principle is shown in Fig. 125 where it can be seen that
the two dies are diametrically opposed and contra-rotating so that the
work can rotate and feed longitudinally between them. Since the form

Fig. 123. Twin die rolling with automatic loading

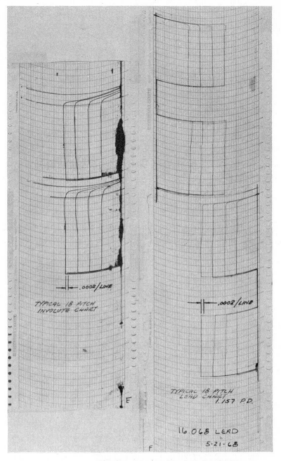

Fig. 124. Graphs of accuracy produced

of the dies is helicoidal (so that they are in fact worms) the process is one of continuous indexing and generating.

While the dies rotate around the axis *aa* they are also oscillated along an elliptical path as shown in the diagram. This tends to lift the dies in and out of engagement and introduces an extremely rapid hammering effect and as in forging this tends to set the material up into a state of plastic flow.

Since the dies are in worm form they are relatively easy to produce accurately – a point of some importance. At the moment the process

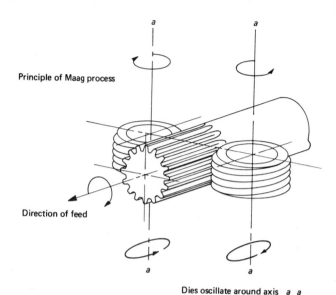

Principle of Maag process

Direction of feed

Dies oscillate around axis *a a*

Fig. 125. Principle of Maag process

appears to be limited to gears up to approximately 125 mm diameter, and helical gears up to 30° helix angle, but it is capable of producing lower pressure angles than the previous process owing to the lower beam strength loading.

Fig. 126. Grob cold rolling machine

The Grob machine is shown in Fig. 126 and, although similar to the Maag, differs in the formation of the dies, which act on a planetary system, as shown in Fig. 127. The two contra-rotating rolling heads are fitted with circular dies mounted around the circumference in a planetary configuration. These small circular dies make intermittent contact with the work so that they contact once every revolution of the main roller head and give the required hammering or impact action.

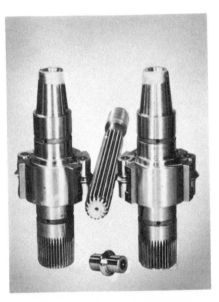

Fig. 127. Grob cold rolling dies working on a planetary system

The process is therefore very similar to the hobbing technique and the necessary synchronization of speeds is obtained from suitable change gears. This technique has the advantage that it can be split into either continuous or interrupted indexing depending on the application.

1.4.5 Grob process (interrupted and continuous indexing)
The indexing drive operates through a geneva mechanism such that at the moment of contact between the component and the roller die the work is stationary. The roller is in contact with the work only while rotating through small angles, therefore it is necessary to stop the work rotating only for this period.

This method is used for producing components with low numbers of teeth, small pressure angle and relatively large depth of tooth, which if generated in a continuous manner would require a wide generating angle α°. Under these circumstances the form of the die would have to be developed to suit and would no longer be an inverse replica of the tooth space of the component.

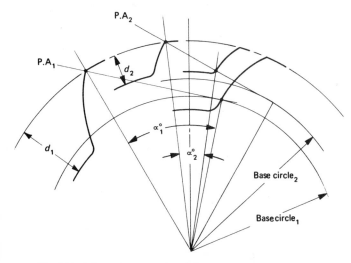

Fig. 128. Effect of depth and pressure angle on generating angle

Fig. 128 shows the condition for both high α_1° and low α_2° generating angles. The latter obviously offers advantages owing to the simpler form of the generating die.

The process is applicable to a variety of forms and is suitable for spur and helical gears, splines, serrations, ratchets and symmetrical or asymmetrical forms. When producing helical gears, however, the machine must be equipped with a differential and either the standard or multiple roller head can be used — the multiple head consisting of a number of roller tools arranged in worm fashion around the die head. The machine is shown in Figs. 129 and 130.

The continuous indexing method is used when producing components requiring a low generating angle and these usually fall in the category of high number of teeth and pressure angle and low tooth depth.

Fig. 129. Rolling helical gears

In this process the rotation of both tool and work is continuous and is not interrupted while the tool is in contact with the work. Since the action is now similar to the generating hobbing process the tool head must be inclined at an angle. The roller heads are inclined at an angle α°. This is found from $\sin \alpha^\circ = \dfrac{P}{\pi DR}$

Where P = pitch of the job
DR = roller diameter

This compensates for the relative displacement between the two units.

Fig. 130. Machine with multiple roller head

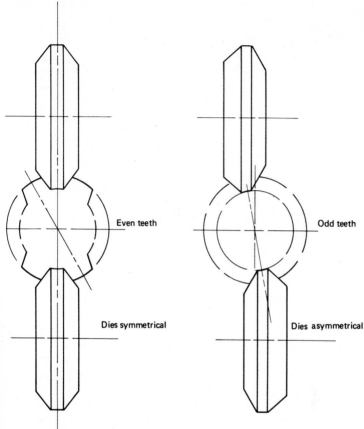

Fig. 131. Die arrangements

The rolling tools are made in high speed steel and are located planetwise in the tool heads with profiles that correspond to the tooth space to be formed.

The arrangement of the dies is as in Fig. 131, which entails either symmetrical or asymmetrical shapes on the tools, depending on whether the number of teeth to be produced is even or uneven. Fig. 132 shows a variety of components which have been produced on the Grob machine.

1.4.6 Blank preparation

The object is to induce the material to grow radially and to limit the growth longitudinally since this can have undesirable side effects. Some

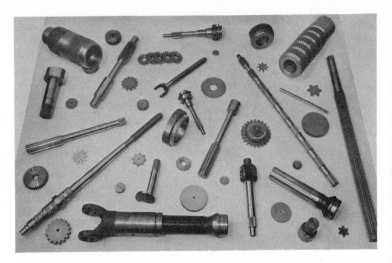

Fig. 132. Components produced on the Grob machine

growth and distortion must be experienced however at both the entry and exit points of the die and the work. Owing to the change in section and working pressures at these points the rate of flow changes, and it is necessary to make due allowance on the shape of pre-machined blanks.

Fig. 133 shows at (a) the condition when one end of the component is closed and at (b) the condition that arises owing to a change in work section. The diagram (c) shows a normal open-ended work section, the effect on the finished gear if no allowance is made on the pre-machined blank, and the modification required to give a parallel finished diameter.

Fig. 134a shows the effect of component shape and pressure angle on the main component forces. The low pressure angle example gives high loading normal to the flanks, low radial or separating forces and high tangential components of force. The high pressure example gives higher radial or separating forces whereas the tangential component and loading normal to the flanks are lower. Fig. 134b shows the effect of component shape on the flow travel which for best results must be kept as low as possible.

In the first example the number of teeth is high and the distance the material is required to move when displacement takes place is reasonably small. The second example is of low number of teeth and low tooth area as compared to tooth space area which entails a high rate of flow travel. The conditions in the second case are undesirable and should be

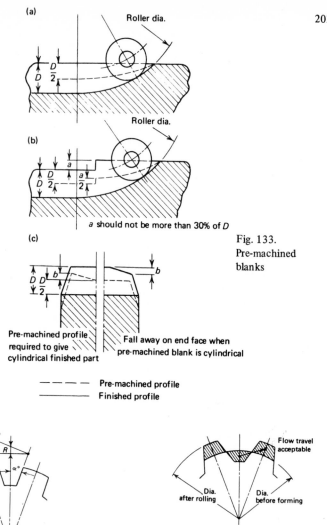

(a) Roller dia.

$\dfrac{D}{2}$

D

(b) Roller dia.

D $\dfrac{D}{2}$ a $\dfrac{a}{2}$

a should not be more than 30% of *D*

(c)

D $\dfrac{D}{2}$ b' b

Pre-machined profile
required to give
cylindrical finished part

Fall away on end face when
pre-machined blank is cylindrical

Fig. 133.
Pre-machined
blanks

– – – – – Pre-machined profile
————— Finished profile

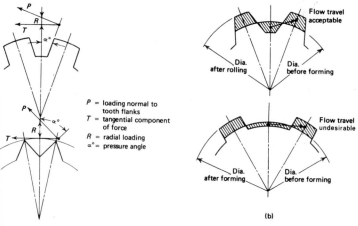

P R T $\alpha°$

P R T $\alpha°$

P = loading normal to
tooth flanks
T = tangential component
of force
R = radial loading
$\alpha°$ = pressure angle

Flow travel
acceptable

Dia.
after rolling

Dia.
before forming

Flow travel
undesirable

Dia.
after forming

Dia.
before forming

(b)

(a)

Fig. 134. Effect of shape on force levels

avoided if possible by doubling the number of teeth and equalizing the tooth/tooth space areas.

1.4.7 Rotary impingement process (Barber Colman)

This is another form of gear rolling which has been patented and developed by the Barber Colman Company in the U.S.A. In this process the forming is done by two identical diametrically opposed dies having profiles conjugate to the tooth space to be produced. The plastic flow and deformation of the metal is obtained by forcing the dies radially inwards by means of cams. The rotation of the die head is synchronized to the work which is continuously rotating and the relative speeds are controlled by suitable change gears.

This radial movement of the dies is performed extremely rapidly and gives the necessary hammering or impact effect necessary to set the material into a state of flow.

In order to form the face width the work is fed axially through the dies which are moved radially in and out once per revolution as they pass over two cams protruding into the bore of the die head housing. Thus the gear tooth spaces are formed progressively around the periphery and parallel to the axes of the blank.

The angular positions of the dies can be changed to suit odd or even numbers of teeth, spur or helical gears, and the pressures exerted are directed into the work whereas reactions are absorbed through a stress frame enclosing the die head.

This process is ideal for bar stock (although individual gear blanks can be produced); the production rate is higher and fewer problems arise from the side effects of axial flow. The bar stock can be parted off into individual blanks of any desired face width, the location for machining being the outside diameter which is produced smooth and concentric by the forming die.

1.4.8 Hot rolling

As distinct from cold rolling, this process has been developed in the U.S.S.R. and to a lesser extent in the U.S.A. for the forming of gear teeth after the blank has suitably been heated. By heating the blank prior to exerting rolling pressure it is easier to get the material into a state of plastic flow. Very little information is available on the process at the time of writing and whether it will ever become commercial.

viable is not known. It is worth mentioning, however, since it is a natural off-shoot of the cold forming process and may have some application in the future.

The process of operating the dies is similar to that employed for cold rolling, two methods being available — (a) using two cylindrical dies at fixed centres, passing the work in bar or stick form between them and (b) for individual blanks when one or both dies are moved inward to create the required pressure.

In the first method the dies are tapered on the leading edges which facilitates the initial engagement of the blank and provides the grip required for accurate deformation. The dies rotate in opposite directions and a timed relationship exists between both dies and the work.

The second method is similar but here the blanks are formed individually and the dies feed radially inwards — end flow is reduced by providing the dies with side plates which protrude outside the addendum of the die to below the root of the gear.

The steel used must be suitable for high frequency heating and in practice only the outer layer of the component is heated to a depth a little greater than the final tooth depth. Induction heating is used to raise the blank temperature to the normal forging range between $850°C - 1000°C$ depending on the material being rolled. As far as the writer is aware the accuracy still leaves something to be desired and the gears must be finished or refined by some other process if any reasonable order of accuracy is required. For some agricultural work or applications where low grade accuracy is acceptable then the gears may be used as rolled.

1.4.9 Summary of rolling methods

Undoubtedly cold rolling made a great deal of progress during the period 1965-68, particularly in the U.S.A. At the time of writing the technique of finishing by rolling is firmly established as a production process in the major organizations producing automatic transmissions and power tools in the States. The technique of finishing by gear shaving is being replaced by rolling, although at present little change has been made in the roughing or semi-finishing operations.

At present the gears are being hobbed or shaped to the same degree of accuracy prior to finishing, but the rolling technique is showing an improvement over shaving in regard to time cycle, consistency and accuracy. It has already been proved that under the right circumstances it is possible to speed up the roughing operation on the gear so that it is virtually a notched blank which will accurately be formed by the rolling dies. There is little doubt in the writer's mind that gears will be rolled from the solid within a few years in the U.S.A., although it must be appreciated that these remarks apply to automatic transmissions and power tools where the pitches are in the range 1¾ - 1¼ mod. and are of simple configuration. There is still much work to be done and problems to be overcome before the larger pitches can be produced from the solid and it remains to be seen if rolling will be as flexible as shaving in respect of small batch work.

Regarding hot rolling, there is not enough information available to enable any reasonable prediction to be made. The rolling technique is developing so rapidly that some of the above data are likely to be out of date by the time this work is printed.

PART 1.5: AUTOMATIC LOADING AND SWARF REMOVAL

1.5.1 Automatic loading

Generating machines have now been developed to such a degree that extremely fast time cycles are often achieved. To ensure better machine utilization and efficiency it becomes necessary to consider means of effecting some form of continuous operation by automatic loading of the machine.

Fig. 135. Automatic loading of cluster gear

The loading mechanism is dependent on the type of machine, the nature of the part, the time cycle and the labour already available. It is not proposed therefore to discuss the best type of loading but merely to point out the types already in use and how they are applied. Briefly the loading mechanisms can be broken down into various types as follows:

(a)   gravity feed units (cheapest and most common);

(b)   magazine loaders and vibratory feeders and hoppers;

(c)   conveyors and arm type loaders   (most sophisticated, used on auto machines and transfer machines).

(a) *Gravity feed*

The main problem here is the control which can be exercised over the parts during loading. Large masses rolling down inclined planes can develop considerable energy and steps have to be taken either to absorb it or stop it from developing. A good example is the cluster gear in Fig. 135 which is both heavy and unstable. In this case a pawl and ratchet assembly is used which lets each component roll by gravity for only a short distance before it is arrested again; this is a form of index conveyor. The offload chute is very short and works entirely by gravity, and owing to the fact that the two rolling diameters are of different sizes side plates are used on the larger diameter in order to prevent one diameter gaining on the other.

Fig. 136. 1B horizontal gear generator, with automatic magazine feed, cutting two 32 teeth involute splines on a mainshaft component

The advantage of this type of loader is that it is cheap, simple, easy to install and relatively easy to change over. Another good example of the gravity type index loader is shown in Fig. 136. Here very long shaft gears are being loaded which are extremely unstable and if loaded by gravity only would tend to move end first. This is really a cascade type loader, the components being indexed one station at a time in cascade fashion by the escapements at each end of the shaft. Gravity is sometimes used to effect the offload since the components are allowed to fall heavy end first to a collecting point, which is dictated by the next sequence to be performed. This particular example is also equipped with a retractable hydraulic steady rest which locates the gear during the cut and prevents deflection.

(b) *Magazine loading and feeders*

Basically the object of the magazine type loader is to provide a storage for the machine to operate automatically for a given period so that one operator can run a number of machines. Fig. 137 shows a good example in that the magazine is versatile and can easily be adapted to suit a number of similar components. The load cycle is extremely short, since it is only necessary to index the turret from one station to the next, and the cycle is continuous until the stations have been completed.

Fig. 137. Automatic turret loading device

An interlock at this point stops the machine from recutting the same gear and a flashing light draws the operator's attention to the fact that the turret needs reloading. If batch quantities increase or operating conditions change so that larger magazine storage becomes necessary, then the turret can be supplemented with a bowl or vibratory type of hopper feed unit. There are a number of proprietary units on the market and they can easily be adapted to give auto load and offload facilities to a turret or arm loader (Fig. 138).

Fig. 138. Hopper feed unit supplementing automatic loading device

Not all magazine loaders are in circular or turret form — there are a variety of configurations which can include gravity feed type, zig-zag storage or even tubular storage, which are ideal for simple disc-type components.

(c) *Conveyors*

The most sophisticated of the auto-loading machines is the conveyor unit with the arm-type loader. This is more common where there is mass production since the loading can be expensive and is not so flexible as the turret loader. The conveyor may be of the shuttle, index or free flow type but the last is the one associated most often with gear cutting requirements. Where production requirements are such that several machines are required to give a certain output per hour, all can be linked to one conveyor which is suitably gated so that components are supplied to each machine. The cut component from each machine is then placed on a gated track on the conveyor and interlocks arranged so that successive machines can unload without fouling components already on the off load track. On the more sophisticated link lines the interlocking required may be extensive and transfer stations are necessary to transfer the gear into the machine loading arms.

Fig. 139. Twin arm loading from free-flow conveyor

Fig. 139 shows a standard type of loader developed by the Sykes organization for conveyor applications. It has the distinct advantage that it can be used on any one of their integrated design machines and is therefore applicable to hobbing, shaping or shaving machines. Components such as those shown have three operations to be performed in

Fig. 140. Integrated line of machines

that the clutch teeth must be shaped and the helical gear hobbed and shaved. This type of component can therefore be placed on a free flow type of conveyor (Fig. 140), the hobber, shaper and shaver units all operating from it with common loading assemblies. The output of the hobber is half that of the other two machines; so to obtain phased output from the line it would be necessary to have two hobber units for each of the others.

The index type of conveyor is sometimes used for certain applications and Fig. 141 shows a shaving machine with an in-line system. The gears are picked up from the conveyor by mechanical arms and once they have been processed they are picked up from the work station and returned to the conveyor whereupon it indexes one station and the cycle is repeated.

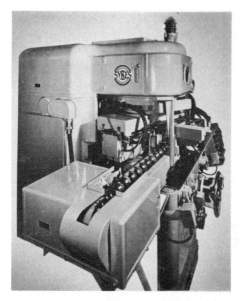

Fig. 141. Automatic loading from index type conveyor

Fig. 142. Composite type automatic loading

A composite type loading method is shown in Fig. 142, where a hobbing and shaping machine are connected together to perform two different operations on the same gear. The hobber is equipped with a type of index conveyor which is in the form of an endless track encircling the machine — the components being ejected automatically after processing so that they slide down a chute by gravity into a vertical storage. The axes of the blanks are turned through 90° during transfer and a hydraulic device pumps the gears up a vertical chute and then across to the shaping machine. They then fall by gravity down another chute and roll along their periphery up to an escapement mechanism. At this point they are released one at a time and are loaded by an air-operated device into the turret mechanism of the shaper. A further device automatically ejects the finished component from the turret where it falls by gravity down the unloading chute.

The Fab-Tec system is now quite widely used for gravity feed-type automatic loading and a typical set-up is illustrated in Fig. 143. The advantage of the system is its flexibility, since it is built up from standard units which are in themselves flexible and easily adjusted. The feed to each machine is by gravity down inclined chutes, and large storage facilities can be provided in spiral elevators. If the blanks are fed through a large number of machines, being gravity feeders

Fig. 143. Fab-Tec system of automatic loading

they can lose considerable height and elevators are then provided to pump them back up to the required height.

1.5.2 Automatic swarf removal

Unfortunately not enough attention is paid to the question of swarf removal. Although the machine tool's principal function is to produce a finished component it also has a by-product — swarf. Attention at the design stage of the machine tool to its removal can save many problems later.

There are two main issues to consider — (a) to remove the swarf from the cutting area swiftly and efficiently, and (b) to filter all oil of swarf swiftly and efficiently and return it to the machine. Obviously the solution to these problems varies with the machine and its application and the following comments are based specifically on machines supplied by the W. E. Sykes company for cutting gears.

On both their vertical and horizontal ranges the emphasis is on ducting the swarf from the cutting area to some other position where it can be handled more conveniently. Fig. 144 shows this in broad

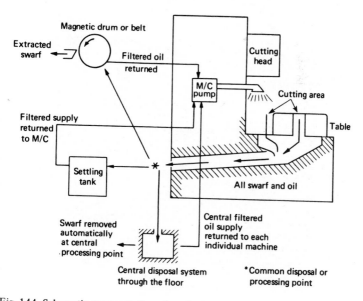

Fig. 144. Schematic representation of methods of swarf removal from cutting oil

principle and the swarf is delivered to the common point * irrespective of how it will be handled or filtered. Having reached this point the issue is how to filter and return the oil to the main machine supply.

To filter the oil effectively requires the removal of the fine particles which are in suspension. One of the best methods is to pass it through a series of settling tanks. Unfortunately this method requires a great deal of room and is not usually possible to put into practice in a machine line.

For this reason the central or common disposal method is useful in that all machines in a line dump into a central channel (usually below the floor) which then takes all the oil away to a common processing area which can be much larger and more efficient than would be possible for each individual machine. The filtered oil is pumped back to each machine individually by means of a separate line. In this manner one large filtering plant is set up to handle all machine tools in a line.

Fig. 145. Automatic swarf extractor on W. E. Sykes I.M.L. range

The disadvantage, of course, is that the disposal channels are fixed and the machines must be operated above them; this limits the flexibility.

A number of proprietary devices are available for extracting the swarf automatically from the cutting oil. Typical of these are the magnetic drum devices. Here the oil and swarf are delivered to the rotating magnetic drum which dips into the contaminated oil. The drum then delivers the swarf to a scraper which removes the particles and cleans the drum. Thus a virtually endless clean magnetic surface is presented for filtering use.

Fig. 145 shows a different magnetic device fitted to one of the Sykes range of machines (the I.M.L.). It is worthy of mention as this swarf removal arrangement is common to all three units — hobber, shaper and shaver. The unit is flanged to the wall of the machine and extends inside it; oil is delivered from the cutting area via a chute on to the extractor. This has a series of magnets on an endless chain running underneath a thin metal cover. The swarf is attracted to the cover by the magnetic field and is caused to climb up the elevator whereby it is delivered outside the machine and removed by a scraper into a suitable receptacle. Fine filters with replaceable cartridges are still necessary to supplement these devices and remove the fine particles which tend to remain suspended in the oil and which if not removed have an adverse effect on tool life.

CUTTING TOOLS

PART 2.1: THE HOB

The conventional hob is made in high speed steel, hardened but not necessarily ground on the form. It can be likened to an involute worm provided with flutes or gashes usually at right angles to the helix and with the tooth profile suitably relieved behind the cutting edges thus formed. A number of threads may be provided but the single start is the most usual. This is preferred for the more accurate work while hobs of two or three starts are used for very high production where the gear is to be finished by some other medium.

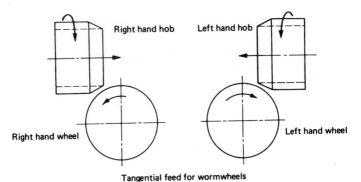

Right hand hob   Left hand hob

Right hand wheel   Left hand wheel

Tangential feed for wormwheels

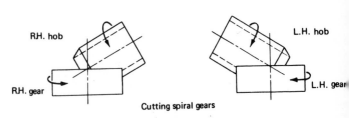

R.H. hob   L.H. hob

R.H. gear   L.H. gear

Cutting spiral gears

Fig. 146. Table indicating directions of rotation and feed when using tapered end hobs

2.1.1 Gear hobs

A taper is usually provided on the leading end of the hob teeth when producing helical gears of 30° helix angle or over. The leading teeth of the hob under these circumstances are very heavily loaded and the tapering or shortening of these teeth eases the load and prevents fracturing. Fig. 146 shows the condition and the table indicates the position of the taper relative to the hand of helix and direction of rotation. It is not necessary to provide such a taper on hobs for spur gears or low helix angle gears.

It is usual for the hand of the hob to be the same as the hand of the gear but hobs of the opposite hand can also be used. On helical gears of 20° helix angle or more it is better for both to be the same hand of helix otherwise the cutting conditions are not ideal and the surface finish produced will be poor.

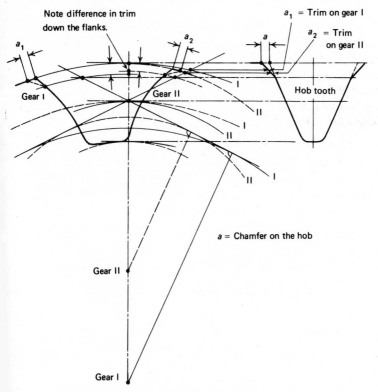

Fig. 147. Variation in the amount of chamfer produced by the same hob on two gears of different numbers of teeth

The basic rack of the hob is usually made in accordance with B.S 2062, Part 1 (1959); the slight modification to the straight side of the rack imparts a small relief to the tip of the gear which allows for deflection of the teeth when under load. The exact amount of relief produced by the hob varies with the number of teeth in the gear — the lower the number of teeth the smaller the relief produced and the higher the number of teeth the larger the relief produced. Fig. 147 shows quite clearly the reason for this variation — the hob would reproduce itself only if it could cut a rack, but the smaller the gear the smaller the radius of curvature and the greater the change of the position of the demand points.

One of the basic requirements of the hob is that it must have the same base pitch as the gear it is producing. In the case of a helical gear this must be in a plane at right angles to the base helix angle.

$$\text{Base pitch} = \frac{\pi}{\text{D.P.}} \times \cos \alpha^\circ,$$

where α° is the pressure angle.

The unusual properties of the involute allow this fact to be exploited because it is not necessary for the meshing pressure angle between the hob and the gear to be the same as the basic rack of the gear. For example the B.S. basic rack is 20° which means that the gear is 20° P.A at the diameter determined from $\frac{\text{N.T.}}{\text{D.P.}}$. If it were required to have a special radius in the root of the teeth of the gear, however, this could be achieved very precisely if the meshing pressure angle between hob and gear were lowered. Fig. 148 shows such a case and the centre of the radius R in the root of the teeth lies on the pitch circle P.D.$_2$.

The base circle diameter of the gear is

$$\frac{\text{N.T.}}{\text{D.P.}} \times \cos \alpha^\circ.$$

and this remains constant however much the meshing pressure angle varies.

The pressure angle at P.D.$_2$

$$\doteq \cos^{-1} \frac{BCD}{\text{P.D.}_2} = \alpha^\circ_2.$$

If the hob were now made with a pressure angle α°_2 it would be necessary to change the pitch in order to maintain the base pitch

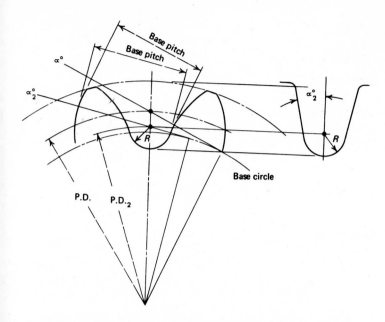

Fig. 148. Changing the meshing pressure angle between the hob and work

Therefore since base pitch

$$= \frac{\pi}{\text{D.P.}} \cos \alpha^{\circ}$$

the D.P.

$$= \frac{\pi \cos \alpha^{\circ}}{\text{base pitch}}$$

Therefore D.P.$_2$

$$= \frac{\pi \times \cos \alpha^{\circ}_2}{\text{base pitch}}$$

$$= \frac{\pi \cos \alpha^{\circ}_2 \times \text{D.P.}}{\pi \cos \alpha^{\circ}}$$

$$= \text{D.P.} \frac{\cos \alpha^{\circ}_2}{\cos \alpha^{\circ}}$$

then D.P.$_2$

$$= \text{D.P.} \frac{\cos \alpha^{\circ}_2}{\cos \alpha^{\circ}}.$$

The hob would therefore be made with a pitch of $\dfrac{\pi}{D.P._2}$ and a pressure angle of α_2 and it would still produce a gear of pitch $\dfrac{\pi}{D.P.}$ and pressure angle $\alpha°$.

The advantage of changing the meshing pressure angle in this case would be the fact that by making the meshing pitch circle go through the centre of the root rad R, the tip of the hob can be made with the same radius and it will reproduce itself exactly in the root of the gear.

2.1.2 Spline and serration hobs

The involute type splines and serrations are virtually stub tooth gears and as such can be treated as gears. There are sometimes fillet or root radius considerations but minimum fillet conditions can usually be met by reducing the meshing pressure angle if necessary, as already described. The straight-sided type of spline or serration can also be generated by hobbing but this requires an entirely different type of treatment.

Whereas with the gear hob the component produced was of involute form and the hob straight-sided, the reverse is true of the spline

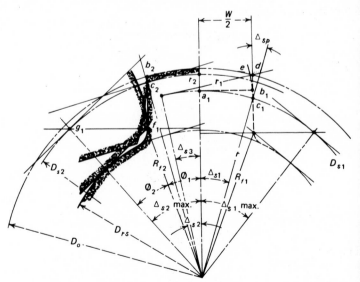

Fig. 149. Generation of straight sided spline forms by hobbing

hob. The usual procedure in developing the hob form for generating parallel splines is to fix the pitch diameter of generation at the outside diameter of the spline. Under these circumstances, however, the fillet generated in the root of the spline may be too large and the fillet can be reduced by designing the hob at a lower generating pitch diameter. Fig. 149 shows the condition for determining the pitch diameter of generation necessary to produce a given fillet in the root.

$$\mathrm{Sin}\,\phi_1{}^\circ = \frac{W}{2R_f}$$

(where W = width of spline key)

$$\text{and } \cos\phi_2{}^\circ = \frac{D_{rs}}{2R_f}$$

$$\text{P.dia. of generation} = \frac{2\sqrt{R_f{}^2 - \dfrac{W^2}{4}}}{\cos(\phi + \phi_2)^\circ}$$

Although this shows the minimum pitch diameter necessary to generate a given fillet the expression does not allow for the outside diameter of the spline which also has a limiting value. The pitch diameter is a function of the outside diameter and spline width and therefore it is necessary to determine if the outside diameter can be achieved with the pitch diameter calculated from the fillet condition.

If reference is made to Fig. 149 it can be seen that for a given spline with pitch diameter D_{s2} the fillet produced in the root of the teeth has a height R_{f_2} from the spline centre. If the pitch diameter is changed to D_{s1} the fillet height is reduced to the value R_{f_1}.

$$R_f = \frac{D_s}{2}\sqrt{1 + \left(\frac{W}{D_s}\right)^2 - \sin^2\triangle_{\max}}$$

and

$$\mathrm{Sin}\,\triangle_{\max} = \tfrac{1}{2}\left[\frac{W}{D_s} + \sqrt{\left(\frac{W}{D_s}\right)^2 + 4D}\,\right]$$

$$\text{where } D = \frac{D_s - D_{rs}}{D_s}$$

In the case of fillet R_{f_1} and pitch diameter D_{s1} however, an adverse condition exists at the outside diameter D_o; the hob has rolled to the point $\triangle_{s1}$ where the flank of the hob now contacts the spline form at point c_1. As the hob continues to roll around the pitch diameter D_{s1} the angle, $\triangle_{s1}$ decreases and point c_1 on the hob describes an involute which has its origin on a base circle equal to D_{s1}. When the angle Δ_{s1} is zero, the point c_1 on the hob has moved to point b_1 and since the arc $a_1 c_1$ is equal to the straight line $a_1 b_1$ the hob is still in contact with the straight flank of the spline. The point c_1 on the hob has not moved in a straight line from c_1 to b_1, but for all practical purposes it can be treated as such since the difference between the arc and the chord over the distance $c_1 b_1$ is exceedingly small. As the hob continues to roll along the pitch circle diameter, however, it reaches the point where the angle Δ_s becomes Δ_{s3}. Since the point is still describing an involute about base circle D_{s1} it has now reached the position e and the instantaneous radius of curvature is r_1. It can be seen from Fig. 149 that the point e has now moved away from the straight flank of the spline and the flank has been trimmed by the amount $(e \text{-} d)$.

If the form of the hob below the pitch line had been carried on at angle Δ_{s1} the flank would have rubbed as it trimmed the tip of the spline. It is therefore desirable to reduce this angle by half while trying to maintain a minimum flank angle on the hob of $5°$ if possible.

Fig. 150 shows the developed hob form necessary to produce the spline in question and the lines $f_1 g_1$, $f_2 g_2$ etc., are the respective lines of action between the hob and the spline for different angles of generation Δ_s. If these lines are extended as shown until they cross, the

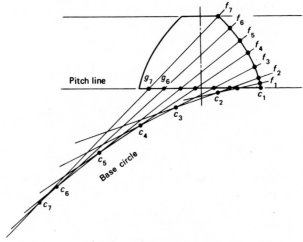

Fig. 150. Development of hob tooth

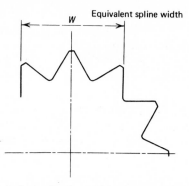

Fig. 151. Equivalent spline key width of a straight sided serration

points c_1, c_2 etc. so formed are the centres of the respective radii of curvature: It becomes equally obvious that the hob tooth form is in fact an involute generated from the base circle formed by the curve $c_1 c_2 - c_6$ etc.

The straight sided serration is treated in a similar manner to the spline; it is only necessary to determine the equivalent spline key width W as shown in Fig. 151 and the calculations are then performed in the same way.

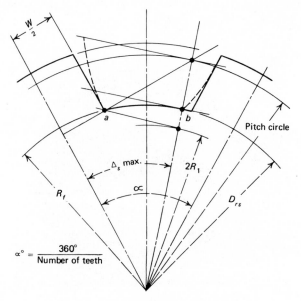

Fig. 152. Generating milling type of spline hob

On occasions it becomes necessary to generate straight sided splines without a radius or fillet in the root of the teeth. It is possible to generate a spline in this manner with a special type of single position hob in which the flanks of the hob generate the spline flanks and the annular groove machined on the tips of the hob teeth mills the root radius of the spline.

Fig. 152 shows the form required and the fillet R_f is now considered as being equal to the root radius D_{rs} and the maximum angle of generation Δ_s is shown as if the hob were generating an imaginary root diameter $2R_1$, where

$$2R_1 = D_{rs}\cos\left(\Delta_s - \sin^{-1}\frac{W}{D_{rs}}\right)$$

The hob teeth are produced in the normal manner but the radius $\dfrac{D_{rs}}{2}$ is produced on the tips of certain teeth as shown between the points a and b. This radius is constant and is produced as an annulus around the hob periphery so that it tends to mill the root of the teeth rather than generate it. The hob (Fig. 153) therefore appears as shown in Fig. 154 and it must accurately be positioned over the component so that the root radius is exactly central. Since the hob is provided with a number of flutes or gashes it is also imperative that the points a and b occur on a flute and not at some intermediate position. The number of

Fig. 153. Single position spline hob

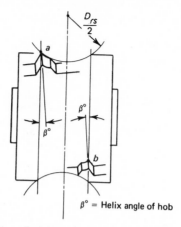

$\beta°$ = Helix angle of hob

Fig. 154. Illustration of critical positioning of single position type hob

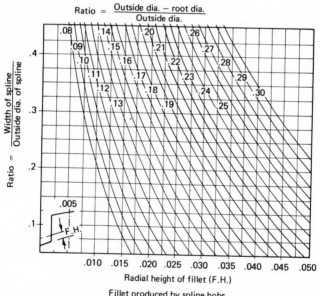

Fillet produced by spline hobs.
Results given by this chart are
only for splines of unit diameter.
For other sizes multiply the fillet
height obtained from the chart by the
outside diameter of the spline.

Fig. 155. Graph indicating fillet height produced by a hob on a given sized spline

flutes required to generate the sharp corners in the root of the spline can be found as follows.

Let F_1 = number of flutes to generate sharp corners

and F = number of flutes in the hob.

Then $F_1 = F \times \dfrac{\beta^\circ}{\alpha^\circ}$ and F_1 must be an integer.

The radius $\dfrac{Drs}{2}$, although an annulus, must be relieved at right angles to the flute face in order that it will retain its position relative to the hob flanks as the hob is sharpened back.

The disadvantage of this type of hob is the fact that it is expensive to produce and requires accurate and skilled setting relative to the component.

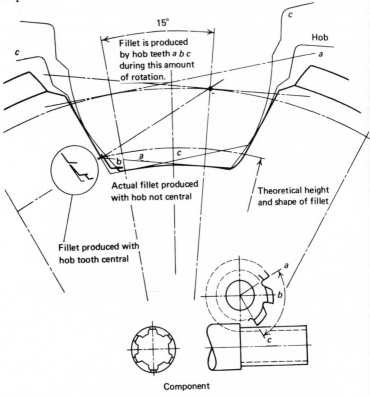

Fig. 156. Errors in spline fillet produced by hobs with insufficient flutes

The fact that the standard type of generating hob produces a fillet in the root of the teeth need not be a problem providing that due allowance is made for it. The graph, Fig. 155, shows the fillet height generated by a hob for a given sized spline. It should be realized that the curve produced is not a radius but a form of trochoid and as such the interference point for all practical purposes lies much closer to the root of the spline.

The most important factor — and one often ignored — is the number of flutes available in the hob to form the fillet. It must be appreciated that the fillet would be formed continuously only if there were an infinite number of flutes in the hob. Since the number of teeth is limited the fillet produced can in fact be larger than the theoretical value, as can be seen in Fig. 156. The fillet is completely formed in this example by a 15° angle of generation and the flutes *a, b, c* do the work. Thus the fillet is produced in three steps instead of a continuous curve. The situation is made more complicated if the hob tooth is not centred before generation starts since then the fillet produced in each corner of the spline is in fact unbalanced and this can form a further fouling point.

The hob shown in the diagram is producing a chamfer on the tips of the spline and is also cutting into an adjacent shoulder which means that the flank must be extended below the hob pitch line in order to form suitable cutting edges. The flank angle is usually made at least 5° in order to avoid rubbing and this means that the shoulder is generated as an involute whose base circle is equal to the spline pitch diameter.

2.1.3 Chain sprocket hobs

These are covered by BS 228 which shows the standard basic rack form of the tool and the type of form produced. The important part of the chain wheel is the radius in the root which receives the roller portion of the chain. The flanks above the roller radius are generated by the 25° flank angle of the hob and as such are involutes from a base circle equal to the sprocket pitch diameter x cos 25°. This type of hob is usually made 'to top' and machines the outside diameter of the blank.

2.1.4 Undercutting and protuberance hobs

Apart from the condition of natural undercut which can be produced by a gear of low number of teeth, the occasion does arise where it is

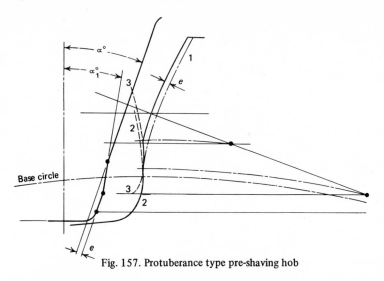

Fig. 157. Protuberance type pre-shaving hob

desirable to undercut the flank by a controlled amount. The gear shaving operation is probably the best example since it is usual to have the shaving allowance on the profile undercut to avoid excessive strain on the tips of the tool. Under these circumstances the tip of the hob is provided with a slight protuberance e usually of the order of about 50-100 microns.

Fig. 157 shows a typical example, the hob tip being carefully proportioned in order to give a smooth and accurate blend of the hobbed fillet with the finished involute profile.

2.1.5 Single position hobs

As the name suggests the design of this kind of tool is such that it can be used only in one position relative to the component and it requires very careful setting. This sort of tool is used primarily for ratchet wheels or similar components which it would be either difficult or impossible to generate in the usual manner. It consists generally of a series of roughing teeth which are used to generate the tooth space of the component to approximately the required shape and these are followed by one finishing tooth which acts as a fly cutter. The finishing tooth is made the same shape as the component tooth space and must accurately be set relative to it. Fig. 158 shows a typical example; it will be noticed that the finishing tooth is offset relative to the component centre line so that the radial flank of the component is angled at $10°$ to the centre line of the hob.

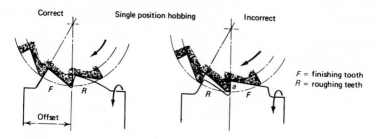

1. Finishing tooth offset to avoid
   flanks rubbing
2. Roughing teeth precede finisher

1. Finishing tooth will rub on flank *a*
2. Roughing teeth follow finisher and have no work to do.

Fig. 158. Correct and incorrect methods of single position hobbing

This gives relief to the tooth flank of the hob tooth and prevents it from rubbing during the cut. Obviously the roughing teeth of the hob must precede the finishing tooth in the cut and therefore the relative directions of rotation are important. In the diagram the view on the right shows the finishing tooth is not offset and the flank will rub

Fig. 159. Flywheel chamfering hob

during the cut; the finishing tooth is shown preceding the roughers and therefore does all the work. Under these circumstances the roughing teeth could be dispensed with. This type of hob is expensive since although all the roughing teeth have the same shape the finishing tooth has to be formed as a separate operation.

A classic example of the single position hob is shown in Fig. 159. This hob is intended for producing the chamfer on starter rings at very high speed. Basically the idea is to make the hob multi-start, usually with the same number of flutes as starts so that one start appears on each flute and all other teeth are removed. Since there is only one tooth of each start on each flute they act as a series of fly cutters and each tooth produces one tooth on the gear. The number of starts is usually six to ten and this gives very rapid indexing of the workpiece and very fast production times.

2.1.6 Wormwheel hobs

To design a hob that will produce a satisfactory wormwheel it is necessary to know both the dimensions of the worm and the method used to produce it. The hob must basically be of the same shape and size as the mating worm but since the hob is form relieved its diameter changes as it is sharpened back. This change in hob diameter has a very big influence on the bearing obtained on the wormwheel and therefore careful control has to be exercised. The change in contact condition is shown in Fig. 160 where it can be seen that when the hob pitch circle diameter is less than that of the worm the bearing is undesirable and the hob should not be used beyond this point.

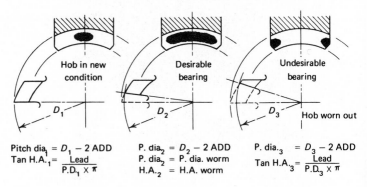

Fig. 160. Contact conditions produced by wormwheel hobs

When the hob pitch circle is larger than that of the worm then its helix angle changes, since the pitch must remain constant. The hob setting angle changes with the hob diameter and the helix angles are only the same when the hob and worm pitch diameters are common.

For extremely accurate applications the form relieved type of hob is used to part-finish the worm wheel only, leaving a small allowance on the flanks to be removed by a shaving hob. This type of hob is not relieved on the form (and cannot be sharpened); consequently the diameter is constant and does not change so that the optimum bearing can be maintained. The life of the hob is extremely limited, as it tends to dull quickly and is suitable only for very small stock removal.

2.1.7 Taper splines

Both the involute and straight-sided type can be produced with a suitable hob. The hob is similar to the conventional type except that it is tapered on its diameter throughout its length. The hob teeth at the starting end are made full depth and they progressively taper to a minimum depth at the finishing end.

In operation the hob is fed tangentially across the component and down the length of the axis at the same time, the centre distance being fixed. The full teeth on the hob cut the full depth of the spline at one end and by the time the hob has traversed through the face width of the component it has also moved across tangentially so that the shallow end of the hob is cutting the shallow part of the spline. Exceptionally accurate leads are required on the hob since it is used over its whole length when cutting any one component; consequently all the cumulative errors in the hob are passed on to the spline.

2.1.8 Inserted tooth and accurate unground hobs

There is a wide difference of opinion regarding the relative merits of the so-called accurate unground hob, the inserted tooth hob and the conventional ground relief hob. On the continent there is wide-spread use of the inserted tooth hob but this is mainly due to the influence of Germany, where some 80% of the hobs used in the car industry are inserted tooth.

In the U.K. only some 5% are inserted tooth, the rest being the conventional type. However the accurate unground kinds are making good progress now that several manufacturers are in a position to supply them. In the U.S.A. the accurate unground type predominates over all others and the inserted tooth has made little headway at the time of writing.

As the name suggests the accurate unground hob is not relief ground after hardening – any growth or unwind that has taken place during heat treatment is not corrected but is kept to a minimum by careful control. Any inaccuracies which arise from distortion are held to an acceptable level by making allowances based on past experience when machining in the soft stage. Hob life is usually better with the unground hob since the ground hob has to be produced by a small diameter wheel, often without coolant, and, due to the rate of wheel wear, a hard grade is necessary which has the tendency to glaze.

High surface temperatures are generated on the hob teeth under these conditions and if not controlled these can affect adversely the metallurgical condition of the teeth. When grinding the form there is also a tendency to produce a small 'heel' at the back of the teeth, as shown in Fig. 161, and this reduces the usable length of the tooth. The amount of 'heel' is also dictated by the relief angle, the higher the relief the larger being the 'heel'.

The unground hob does not suffer from this limitation and it is possible to provide higher relief angles and a longer usable length of tooth.

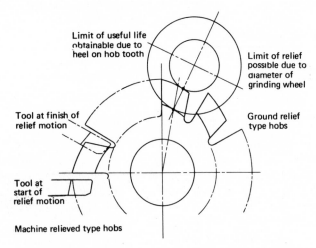

Fig. 161. Comparison of ground and machine relieved type hobs

The inserted tooth arrangement is built up of separate blades held into a body by rings shrunk on the ends. The blades are usually ground in a separate fixture on a worm grinding machine, enabling a large wheel to be used with coolant and producing no heel. The cutting relief is obtained by the manner in which the blades are mounted in the body, so that the length of usable tooth is similar to that obtained in the unground type. The blades are made and heat treated individually before insertion in the body which enables closer control in heat treatment to be obtained and a slightly better metallurgical structure.

PART 2.2: SHAPER CUTTERS

2.2.1 Types of tool

Sometimes referred to as the pinion type cutter, the shaper cutter
made in the form of a gear with cutting edges and is available in
number of forms and variations depending on the nature and shape c
the workpiece.

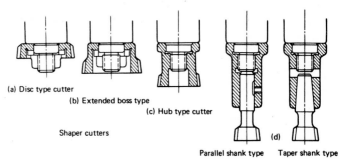

(a) Disc type cutter

(b) Extended boss type

(c) Hub type cutter

Shaper cutters

(d)

Parallel shank type   Taper shank type

Fig. 162. Various types of shaper cutter

Some form of addendum correction is usually applied to the tool i
order to permit the best conditions after resharpening. The geometry c
the tooth form can be seen from Fig. 163 where the cutter is develope
as a ribbon rack — the top rake angle T gives the necessary clearan
on the flanks to avoid rubbing and the front rake angle F is to assi
cutting.

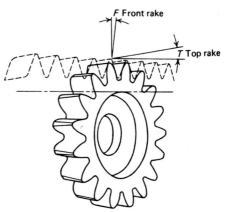

Fig. 163. The cutter and its ribbon rack

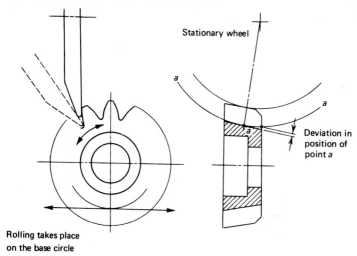

Fig. 164. Stationary wheel type grinding

The teeth on the cutter are ground on a generating grinding machine where the flank of the wheel represents one or both sides of the basic rack depending on the method employed.

In the past the most common method used the stationary grinding wheel which produced one flank at a time as shown in Fig. 164. The results obtained are satisfactory providing only true involute profiles are required, but if modifications to the involute are essential then there are certain limitations to the process. The wheel represents one flank of a rack at zero pressure angle; consequently the tip of the wheel follows the path *a* from front to back of the cutter. It is obvious from the diagram that if a modification is introduced to the straight side of the wheel then the position of this modification varies from front to back of the tool. This means that although modifications may be introduced within reason, they cannot accurately be reproduced by the cutter throughout its life.

The best method of producing the cutter is with a reciprocating grinding wheel as shown in Fig. 165. The wheel now uses both flanks dressed at the generating pressure angle and if profile modifications are imparted to the wheel they are transferred accurately to the cutter at the correct position owing to its reciprocating action. Such a cutter will produce a constant basic rack throughout its life.

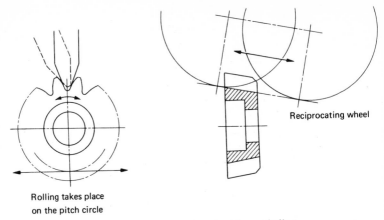

Reciprocating wheel

Rolling takes place
on the pitch circle

Fig. 165. Reciprocating wheel type grinding

2.2.2 Helical cutters

These are available in the same configurations as spur type cutters but have an added complication due to the helix angle. Fig. 166 shows a cutter with its developed lead and this must always be equal to the lead of the guide used to control it. Fig. 167 shows the two types of sharpening for helical cutters — (a) the normally sharpened cutter, where each tooth is sharpened at right angles to the helix, and (b) the Sykes sharpened cutter where the cutting edges are lip sharpened.

The latter method of sharpening must always be used when producing continuous double helical type gears or the gap type where the gap is exceedingly small. The former is used for most other applications as it is generally accepted that it gives optimum durability and productivity. For high accuracy applications the Sykes method of sharpening

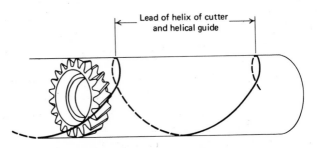

Lead of helix of cutter
and helical guide

Fig. 166. Helical cutter and its developed lead

(a)

Fig. 167. Two types of
sharpening for helical cutters

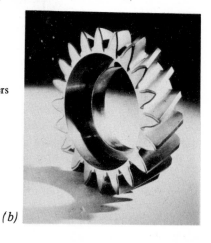

(b)

should be used since it does not distort the cutting edge. A chamfer is
provided on the acute side and a groove on the obtuse side and both
are ground from the base circle of the cutter so that they follow the
involute profile exactly. The normally sharpened cutter should not be
used for high accuracy finishing operations since the method of

sharpening distorts the cutting edges and causes slight profile errors.
Theoretically the cutting face should be curved since it should be
normal to the helix at each diameter on the cutter. The helix angle at
each diameter changes, therefore the front face should in fact follow a
form similar to a radial helicoid. It is virtually impossible in practice to
sharpen such a profile on the cutter since the overall shape varies
relative to the outside diameter. Owing to this distortion the profile of
the envelope of the cutting edges varies as the cutter is sharpened but
the errors are of acceptable proportions on cutters up to 10 D.P. 30°
H.A. In practice the cutters are sharpened normal to the helix angle at
the pitch circle diameter and with the axis tilted to the front rake
angle.

2.2.3 Spline and serration cutters

These can be manufactured for straight-sided splines and serration
but as with the hob there are a number of limitations to the process.
The hob has the advantage that once the basic rack necessary to generate
a given form has been determined, this can accurately be maintained as
the hob is sharpened back and does not change, for all practical purposes, with the diameter of the tool. The cutter, however, is not a rack
and therefore the meshing conditions between the cutter and the component change as the tool is sharpened. It is necessary for the basic rack
of the cutter to be constant from front to back if the cutter is accurately
to be reproduced. If the basic rack of the cutter changes with its diameter, the problem arises as to how to produce the cutter form
accurately. Under these circumstances it is necessary to find the form
required on the cutter in both the new and worn conditions, then, by
using a formed wheel at a given angle, to grind the flanks; finally
using a second formed wheel at a different angle, the diameter is ground.

It can be seen that this requires both accurate setting of the wheel
relative to the work and accurate dressing of the respective wheels. The
checking of the cutter is carried out optically and this means that it is
only practical to check the form on the cutting edge. Under these
circumstances most manufacturers of cutters will undertake this sort of
work only if the component tolerances are reasonably wide and the
cutter life can be restricted.

If the basic rack at the front and back of the cutter is the same
then the grinding wheel can be dressed to this form and the tool can
accurately be generated. Fig. 168 shows the meshing conditions when a
gear shaper cutter is used for generating internal straight-sided splines.

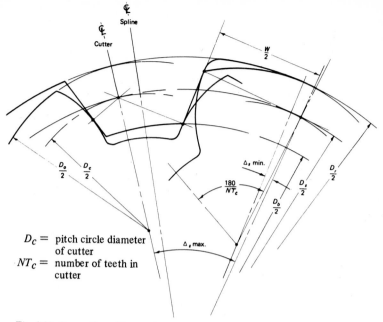

D_c = pitch circle diameter of cutter

NT_c = number of teeth in cutter

Fig. 168. Generation of internal straight sided splines by a gear shaper cutter

Maximum angle of generation = Δ_s

$$\cos \Delta_s = \frac{\sqrt{R_b{}^2 - \dfrac{W^2}{4}}}{\dfrac{D_s}{2}}$$

Where D_s = p.c.dia.
R_b = bore radius
R_f = fillet radius

and $R_f = \dfrac{D_s}{2} \sqrt{1 + \left(\dfrac{W}{D_s}\right)^2 - \sin^2 \Delta_s \text{ min.}}$

The geometry is greatly simplified if the cutter can be manufactured with half the number of teeth in the spline or serration. Figs. 169 and 170 show the conditions, so it can be seen that the profile of the cutter is now a function of the equivalent spline width.

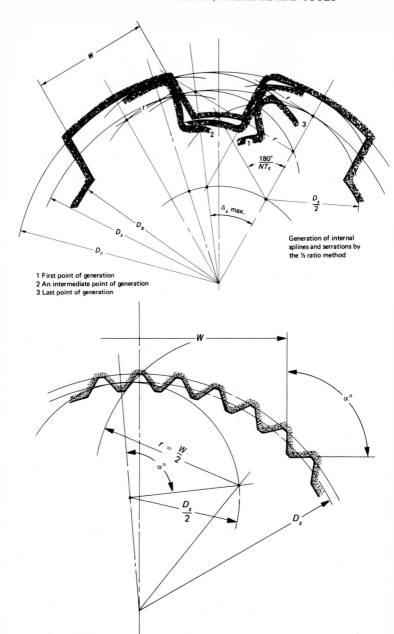

Generation of internal
splines and serrations by
the ½ ratio method

1 First point of generation
2 An intermediate point of generation
3 Last point of generation

Figs. 169 and 170. Generation of internal splines and serrations by the
half ratio method

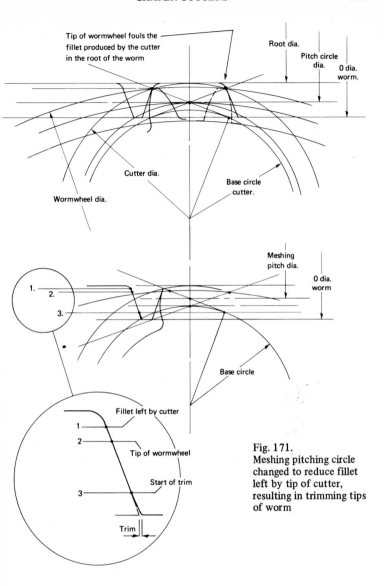

Fig. 171.
Meshing pitching circle
changed to reduce fillet
left by tip of cutter,
resulting in trimming tips
of worm

2.2.4 Worm generating cutters

This type of tool is used on a hobbing machine where the cutting action is reversed — the worm to be produced is placed on the hob arbor and the cutter on the work arbor. The cutting edges must lie on the centre line of the work and the component produced should be reasonably small pitch and low helix angle for best results. Owing to tl large number of teeth in action at any one instant the chip loading high. Care must be exercised in the use of this type of cutter meshing conditions offer severe limitation, as may be seen in Fig. 17 The cutter itself is small relative to the mating wormwheel a⬛ the fillet produced by the cutter in the root of the worm must cle the addendum of the wheel. The fillet may be reduced by lowering tl meshing pitch diameter but this in turn brings the base circle of tl cutter inside the addendum of the worm and therefore trims away son of the profile.

One of the problems encountered with the shaper cutter is ⬛ tendency to cut a different root diameter compared to a given too⬛ thickness throughout its life. Fortunately the change is relatively sm⬛ and for most applications it may be ignored. However, there are sor instances, such as gear pump rotors, where the root diameter and too⬛ thickness are required accurately to be controlled. In such cases gre care has to be exercised in the design of the cutter in order to keep t⬛ change to a minimum. The reason for this apparent change in too⬛ size may be seen from Fig. 172. The cutter is shown in three conditio⬛ — (a) when new and enlarged on diameter the meshing pressure an⬛ between cutter and gear = $\alpha_1°$; (b) nominal diameter and meshi⬛ pressure angle $\alpha°$; (c) when worn out and reduced on diameter, meshi⬛ pressure angle $\alpha_2°$. The meshing pressure angle between the basic ra⬛ and the cutter and the basic rack and the gear remains constant; in t⬛ case of the cutter only the point of contact between the flank of t⬛ cutter and rack changes. It can be seen from the first and last diagra⬛ that although the addendum of the cutter is contacting the root d⬛ meter at all times the flanks of the cutter and gear are not in cont⬛ and that a backlash a of several thousands of an inch is present. As t⬛ tooth thickness produced is usually the means of sizing the gear t⬛ cutter would be fed further into depth until the correct thickness w⬛ obtained and thus the root diameter would be produced undersize.

2.2.5 Cutters for internal gears

In addition to the limitations already described there are furth⬛ physical limits which although obvious are often overlooked. The fa⬛ width of an external gear relative to its diameter is not a limitation, f⬛ provided that the machine in use has sufficient capacity, it does n⬛ affect the design of the cutter. It does, however, affect the cutter desi⬛

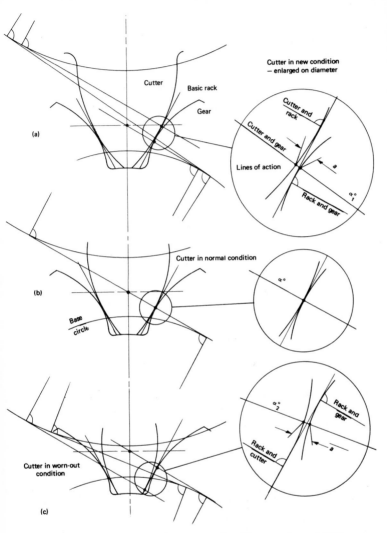

Fig. 172. Variation in root diameter produced by gear shaper cutter

if the gear or part to be produced is internal, since the cutter must be smaller in diameter than the part it is making. This point is particularly important when producing small internal serrations since the cutter is usually made half the diameter of the component. Fig. 173 shows an internal serration 25mm diameter, 40mm face width, being produced

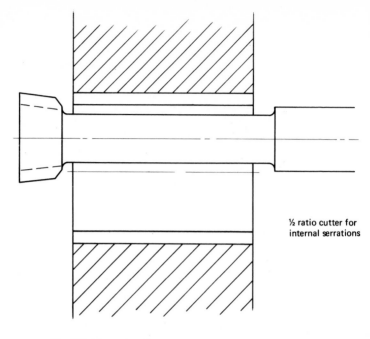

½ ratio cutter for
internal serrations

Fig. 173. Example of impractical application of serration cutter

by a ½ ratio cutter. It can be seen that the application is not ve
practical because (a) the manufacture of the cutter is extremely difficu
and (b) the deflection of the cutter when under load will be conside
able. Care must be exercised, therefore, with internal applications
see that the face width to diameter ratio is reasonable.

The limitations of the shaping process for internal gears have alread
been discussed on pages 85-90. It has been pointed out that there is
limit to the number of teeth which can be used in a cutter to generate
given internal gear. This limitation is a function of the tooth dept
P.A. and diameter of the gear. The graph (Fig. 174) shows the limiti
factors for number of teeth in cutter relative to the gear for differe
pressure angles, assuming full depth, i.e., 2.25 × mod. If the too
depth is decreased more teeth can be provided in the cutter withou
interference. It must be emphasized, however, that this allows for on
some of the forms of interference which may take place and referen
should be made to pages 85-90 as already mentioned.

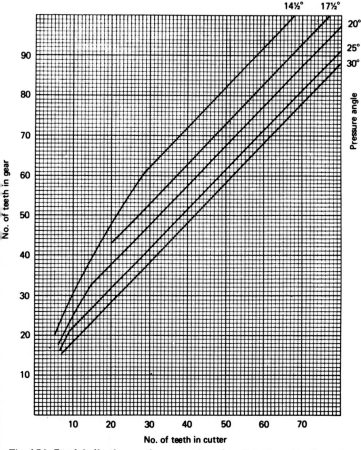

Fig. 174. Graph indicating maximum number of teeth in the cutter for a given internal gear

.2.6 Chain sprocket cutters

This type of component can also be produced with a gear shaper utter. The flanks are involutes and the tips are generated with a adius to suit the roller of the chain. Since accurate blending of this tip adius with the involute is essential and must be maintained as the utter is sharpened back, the cutter is best produced by the 'thru-grind' rocess already described. The limitation with this process, however, is he pitch of the chain since on pitches of say 37mm or larger the rinding wheel width required is excessive. Apart from this manufacturing

Fig. 175. A pair of 37 mm pitch Syhypro chain sprocket cutters

limitation the large area of the cutter profile means that the ch
load when cutting is also high. It becomes desirable, therefore,
reduce the face area of the cutter presented to the work in order
minimize the size of the developed chip. W. E. Sykes have patented
method for producing large chain pitches which overcomes this proble
and which also gives excellent productivity. The cutters are sold und
the trade name 'Syhypro' and are clearly shown in Fig. 175. The proce
can be used only on the Sykes type twin spindle machines as t
essential part of the process is that the cutters are made in pairs su
that one tool is used on each spindle. Both tools operate on the sar
tooth space; one tool produces the left-hand flank and the other t
right-hand flank with a suitable overlap at the centre so that the roll
radius is produced with a smooth blend. In this manner the area of t
profile of the cutter to be generated is reduced thus minimizing t
width of the grinding wheel required. Also the size of the develop
chip produced by the cutter is smaller since each cutter cuts alternate
Owing to the manner in which the chip area is broken down excelle
productivity results and much higher feed rates and deeper cuts can
taken than with the conventional type of tool.

2.2.7 Special form cutters

The generating process is not limited to involute forms but a varie
of profiles may be produced providing certain limitations are borne
mind. The top angle of the cutter must be kept to a minimum (usua

°) in order to limit the change in diameter of the cutter from the new o the worn-out condition. In order to produce the cutter it must either e conjugate to the same basic rack from front to back of its face width r have a constant profile. The latter requirement applies to forms vhich cannot be generated and which are produced with a formed vheel. Under these circumstances the cutter is produced with the ame form from front to back. It is fairly obvious that in this event the orm produced on the component will vary slightly as the cutter is educed in diameter, and for this reason the diameter change must be ept as small as possible.

Providing slight changes in the required profile can be permitted hrough the life of the cutter, a variety of shapes and configurations can e produced rapidly and economically. Figs. 176a, b and c show some xcellent examples of what may be achieved.

Sometimes the profile to be produced is made up of a portion vhich can be generated and a portion which must be regarded as pecial' and following a different law. Fig. 177 shows a good example here the sprocket teeth, although elliptical, are considered as part of larger sprocket having a centre O and pitch diameter D_s, so the form f the cutter is known and can be generated from a constant basic ack. The shape on the end of the sprocket, however, follows a different orm and is built up from a radius r struck from the major axis of the llipse at a distance a from the centre line OO. The form of the cutter equired to generate this portion of the sprocket can be calculated as ollows.

ig. 176a. Cutter with replica of cam profile for 'back generating' profile of gear shaper cutter

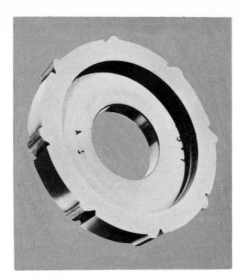

Fig. 176b.
Cutter for special
ratchet profile

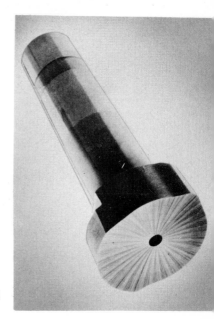

Fig. 176c.
Shank type special
form cutter

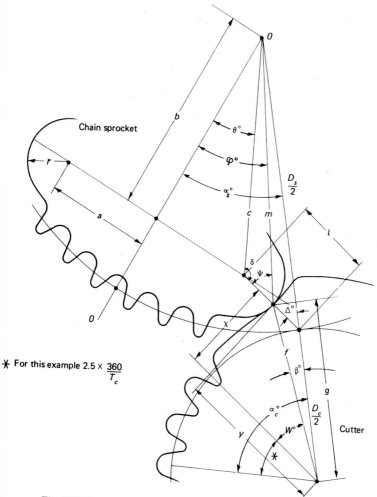

Fig. 177. Special gear shaper cutter for elliptical chain sprocket

Let T_s = number of teeth in sprocket;
D_s = pitch diameter of sprocket;
D_c = pitch diameter of cutter; and
T_c = number of teeth in cutter.

Now choose a number of values of φ° between the maximum and minimum and calculate the x and y co-ordinates of the cutter profile relative to the nearest centre line of the cutter tooth space, where

$$\varphi^{\circ}{}_{min} = \theta^{\circ} = \alpha_s{}^{\circ} = \tan^{-1} \frac{a}{b},$$

$$\text{and } \varphi^{\circ}{}_{max} = \tan^{-1} \frac{a+r}{b}$$

Now $\alpha_c{}^{\circ} = \alpha_s{}^{\circ} \times \dfrac{T_s}{T_c}$

then $\sin \psi^{\circ} = \dfrac{c}{r} \sin (\varphi^{\circ} - \theta^{\circ})$,

and $m = \dfrac{r \sin \delta^{\circ}}{\sin (\varphi^{\circ} - \theta^{\circ})}$,

where $\delta^{\circ} = 180^{\circ} - [\psi^{\circ} + (\varphi^{\circ} - \theta^{\circ})]$.

Also $\sin \Delta^{\circ} = \dfrac{C \sin \delta^{\circ}}{\frac{1}{2} D_s}$,

$\alpha_s{}^{\circ} - \theta^{\circ} = 180^{\circ} - (\delta + \Delta^{\circ})$;

$\therefore \alpha_s{}^{\circ} = 180^{\circ} - (\delta^{\circ} + \Delta^{\circ}) + \theta^{\circ}$.

$g = \text{centres} - m \cos (\alpha_s - \varphi^{\circ})$

$\tan \beta^{\circ} = \dfrac{m \sin (\alpha_s - \varphi^{\circ})}{g}$,

$\omega^{\circ} = \alpha_c{}^{\circ} - \beta^{\circ} - \left(2.5 \times \dfrac{360^{\circ}}{T_c}\right)$

$f = g / \cos \beta^{\circ}$.

Therefore $x = f \sin \omega^{\circ}$ and $y = f \cos \omega^{\circ}$.

2.2.8 Special profile modifications

We have already discussed the two grinding processes for producing gear shaper cutters and the question of profile modifications. These deviations from the true involute profile are usually in the order of 5 to 25 microns but there are other types of modification used for chamfering the tips of the gear teeth which may be in the order of 0.12 to 1.25mm.

The cutter produced by the stationary wheel process is very limited for this order of magnification whereas the cutters produced by the reciprocating wheel process give much more life. This technique was patented by the Sykes organization and marketed by them as the 'Thrugrind' cutter; the following example shows the increased tool life which is possible relative to the conventional method of grinding.

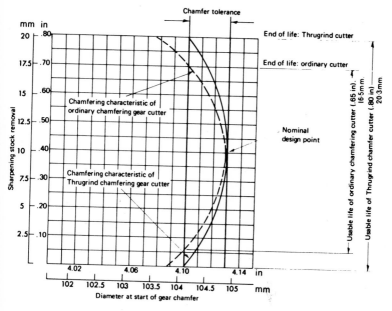

Fig. 178. Comparison of chamfer produced by different cutters

Fig. 178 illustrates a comparison between the two processes when both are designed to produce the same degree of chamfer. In each case the ability of the cutter to produce tip chamfer will change as the cutter is sharpened back through the face width. The change is from minimum chamfer, progressing to the nominal design point, back to minimum chamfer and finally out of tolerance on the low side. Although both cutters are capable of producing the tip chamfer within the required tolerance, the chamfer produced by the conventional cutter goes out of tolerance after 16mm of sharpening stock has been removed. The 'Thrugrind' cutter is still producing acceptable work after 20mm has been removed from the face width. Since the shaping process inherently involves a change in any involute modification on the gear teeth as the cutter is sharpened back, the optimum conditions will exist when the cutter will produce the required modification within a desired tolerance through as many sharpenings as possible.

PART 2.3: SHAVING TOOLS

The shaving process with its advantages and limitations has already been dealt with in some detail in 1.3.1 (page 129). The following remarks therefore relate specifically to the tool. Fig. 179 shows a variety of shaving tools; the cut is obtained from the serrations provided down the flanks. These are not perfect in configuration because of the difficulty in producing them and very little has been published

Fig. 179. Various shaving tools

regarding the best configuration. This is possibly due to (a) the number of variables involved in practical tests; (b) manufacturers regarding this type of information as confidential, and (c) the practical difficulties involved in producing variations of the serration profile. The form shown in Fig. 180 (a) can be produced without difficulty and is the configuration most commonly used. The variant (b) would appear to be a better configuration since both cutting edges would then lie normal to the helix angle. However this is difficult to produce since it is not possible to provide any relief to the flanks of the slotting tool used to produce them. Suggestions have been made that the best configuration for the serrations would be as shown at (c), where they lie diagonally across the flanks and thus follow the effective path of contact between the flank of the tool and gear. This would have the effect of allowing for the present variation in cut experienced due to the cutting edges sweeping diagonally across the flanks of the gear. This is difficult to

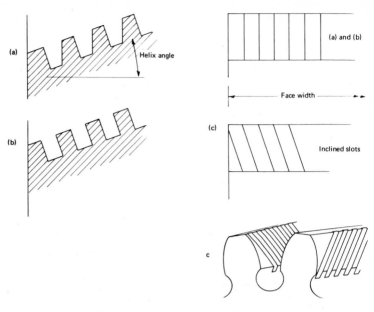

Fig. 180. Shaving tool-slot configurations

prove due to the practical limitations involved in producing such a configuration on the shaving tool.

For the reasons already disclosed the shaving tool should be designed for a specific gear but since this is not always possible on economic grounds it becomes necessary to consider what is involved in attempting to widen the range. The first consideration is the hand and helix angle of both tool and gear. The hand of the tool should be opposite to that of the work and the crossed axes angle should be limited between 5° and 15°.

It follows, therefore, that by intelligent choice beforehand a range of tools could be devized to take advantage of these limits and the following chart serves as a good guide.

Tool	Spur	15° R.H.	15° L.H.	25° R.H.	25° L.H.
H.A. range (°)	5 - 15	0 - 10 20 - 30	0 - 10 20 - 30	10 - 20 30 - 40	10 - 20 30 - 40
Hand of gear	right and left	left	right	left	right

D.P.	4	5	6 & 7	8 & 9	10 & 11	12 to 15	15 to 19
Mod.	6	5	4 & 3.5	3 & 2.75	2.5 & 2.25	2 to 1.7	1.7 to 1.3
P.C.D. of tool and tooth range produced							
8½ in (220 mm)	18–22	18–23	18–24	18–25	18–26	18–27	18–28
8 in (200 mm)	22–27	23–29	24–31	25–34	26–37	27–40	28–43
	27–32	29–36	31–41	34–48	37–55	40–65	43–77
7½ in (190 mm)	32–40	36–48	41–58	48–72	55–99		
	40–50	48–65	58–90				
	50–65	65–99					
	65–90						
7 in (180 mm)							

Fig. 181. Table indicating possible range of teeth produced by one shaving tool

The second consideration is more difficult since this relates to the range of teeth which can be covered by a given tool, this in turn depending to a certain extent on the helix angle of the work. It should be understood that this is not a physical limitation since a tool can be used to shave any number of teeth but if the range is exceeded a gradual deterioration takes place in the quality of involute profile produced. This deterioration becomes acute on gears of large pitch and low number of teeth. The choice of range is therefore difficult in that opinions and requirements vary as to what constitutes an acceptable variation. The table (Fig. 181) is therefore only a general guide to the nominal range that one tool will cover. It shows the range of teeth of standard uncorrected 20° P.A. spur gears over which normal commercial profile accuracy can be attained. The choice is particularly difficult in the areas below thirty teeth shown shaded and in these cases most tool manufacturers will request details of the mating parts since these have a big influence on the design of the tool.

PART 2.4: MILLING CUTTERS

The milling of gears using suitably form relieved milling cutters is a dying art since it has severe limitations regarding the accuracy produced and is not as fast as the generating process which superseded it. The cutters were designed to cover a range of teeth and since this is a form milling process the inherent inaccuracy in the cutter profile is obvious.

The multi-form annular grooved milling cutter is still a very useful production tool for profiles which cannot be generated. A typical example of a modern cutter is shown in Fig. 182 – a milling cutter used for the production of the 'fir tree root' on turbine blades. Multi-form cutters are also used for producing pulverizing knives for the food processing industry. The flutes are milled with a spiral; the object being to reduce the instantaneous tooth load and give a smoother cut. The helix angle should be large enough to give a helical overlap of the flute pitch in order to obtain optimum smoothness in operation.

The larger profiles give problems in regard to the volume of chips produced – the helical flute improves the situation but does not eliminate it in the case of deep sections. A further improvement may be effected under these circumstances by providing chip breaker grooves. These break up the chips and help to avoid overloading and the packing of the chips between the flutes in the cut. Again, to alleviate the tooth

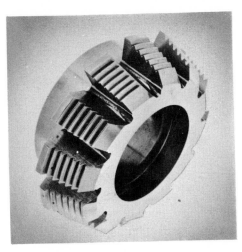

Fig. 182. Special form milling cutter

load on large or deep sections, the cutters can be made with staggered flutes — the form is produced alternately by the right and left hand flutes. Thus only half the profile is produced by any single flute, so reducing the load.

PART 2.5: TOOL SHARPENING

2.5.1 Shaper cutters

The sharpening of a gear shaper cutter is an operation requiring care and attention if the original precision of the tool is to be maintained throughout its life. The suitability of the grinding machine and any fixtures employed has an undoubted effect on the quality of sharpening. Both should be sufficiently robust to accommodate the cutter without vibration or deflection and the grinding machine must be in good condition for best results.

Spur or cone sharpened cutters are sharpened by grinding the front cutting face. Since the cutter is provided with a top rake it follows that grinding this face reduces the cutter diameter. It is essential, therefore, that the grinding is uniform since lack of uniformity causes eccentricity of the pitch circle with resultant variations in tooth thickness. The front rake angle, which is usually 5°, controls the form of the cutting profile and must therefore be maintained as accurately as possible. Sharpening can take place on a rotary surface grinder with an inclined magnetic table set to the front rake of the cutter. Dependent on the condition of the cutter it may be necessary to rough and finish but it is good practice to sharpen the cutter before it becomes too worn. The worn area on the cutter should be such that it could be removed by grinding between 0·13 and 0.4mm from the front face depending on the pitch. Grinding wheel diameters vary between 125 and 300mm depending on the application and a vitrified bond of 60 GRIT J Grade is suitable for average conditions. Light, even cuts should be taken using plenty of coolant and great care must be taken to avoid overheating which tempers the cutting edge.

(a) *Normally sharpened type*

This type of sharpening requires the use of a special attachment since the cutting edges lie in planes normal to the helix angle at the pitch line. The attachment shown in Fig. 183 can be used on a normal tool and cutter grinder or any grinder which has facility for lateral table movement. It consists of an accurately set base inclined at 5° to produce the required face angle while spring-loaded plungers and interchangeable cutter locations provide the means of indexing and locating the cutter during sharpening. For most applications a 100 to 125mm diameter wheel, 46 Grit K grade for roughing and 60 Grit for finishing are suitable.

Fig. 183. Sharpening attachment for helical cutters

(b) *Sykes sharpened type*

This type of sharpening is essential if continuous double helical gears are being produced. Therefore the situation can be complicated by the fact that the cutters are usually supplied in pairs of opposite hand. This is not always so, however, since this type of sharpening is also used on single helical gears where extreme accuracy is required. The faces are first ground square to the axis of the cutter and where paired cutters are involved care should be taken to see that the diameters are matched within 25 microns. Fig. 184 shows the Sykes type sharpening

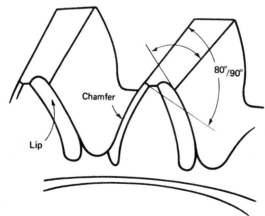

Fig. 184. Sykes sharpened cutter for continuous double helical gears

where a slight chamfer is provided on the side having excessive relief while a lip is formed on the side having negative relief. Sykes manufacture a special attachment shown in Fig. 185 which allows the grinding wheel to follow accurately the involute profile. The attachment can be used on any tool and cutter grinder provided the wheel head can be used with an extension spindle as shown. The wheel must be dressed to the form shown in the diagram since the shape of both the lip and the chamfer is important if optimum results are to be achieved.

Fig. 185. Sykes type sharpening attachment

Grinding wheels of 40mm diameter with widths of between 6mm and 12mm depending on the pitch are recommended, and these should be K grade, 80-100 grit size. Owing to the small diameter wheel necessary it is sometimes difficult to obtain sufficient spindle speed to give the peripheral speeds of the order of 1800 metres/min necessary for efficient grinding, but every effort should be made to keep this as high as possible. Very light even cuts with frequent wheel dressing are essential, since grinding with this attachment takes place without coolant and great care is necessary to avoid burning, particularly at the tips where the lip and chamfer meet.

(c) *Machines*

Fig. 186 shows an automatic cutter sharpening machine suitable for sharpening spur type cutters and helical cutters of the 'normally sharpened' type. The work slide is equipped with twin tables, one table arranged permanently for spur type tools as in Fig. 187, and the other with an attachment for helical cutters which can be seen in Fig. 186.

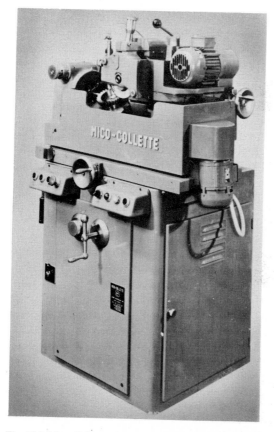

Fig. 186. Mico-Collette automatic cutter sharpening machine

The work slide reciprocates past the wheel head, both length of stroke and feed rate being adjustable. The wheel head is adjustable through 90° between two fixed stops, the axis of the wheel lying at right angles to the table movement when producing helical tools and parallel to the table movement when producing spur tools. Cutter with Sykes type sharpening cannot be sharpened on this machine and require an attachment as previously described.

Fig. 187. Mico-Collette automatic cutter sharpening machine, showing one table arranged permanently for spur type tools

(d) *Effect of errors*

It often occurs that a new cutter produces perfect gears but ceases to do so after sharpening by the user. The sharpening itself is a relatively simple operation but care must be exercised to ensure that the tool is running true and is accurately set relative to the grinding wheel. The standard front rake angle is usually $5°$ and is shown as α in Fig. 188 which indicates quite clearly the effect of changing the face angle to $\alpha_{i1}°$. The effect is to deepen the cutter tooth from d to d_1 thus changing the pressure angle from $\varphi°$ to $\varphi_1°$.

If the cutters are eccentric to the axis of rotation XX (Fig. 189) (or are running out on the back face A), the cutting face becomes biased and has local high and low spots which in turn leave some teeth high relative to the others. This has the combined effect of P.C.D. eccentricity and adjacent pitch error but this again can further be complicated when sharpening helical type cutters, since each tooth face is sharpened individually. Obviously accurate indexing from tooth to tooth when sharpening and positioning of the wheel relative to the cutter are critical. The cutters with teeth sharpened normal to the helix require careful setting of the helix angle on each tooth face since errors in this angle distort the involute profile on the cutting edge.

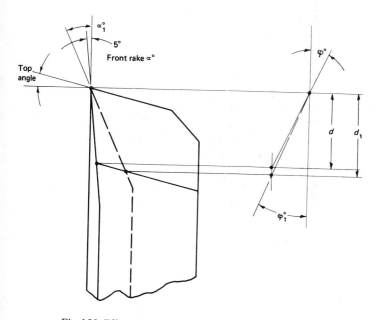

Fig. 188. Effect of changing face angle on shaper cutters

Eccentricity on the back face of the cutter with this type of sharpening can cause pitch errors in the tool as shown in Fig. 189. Similar effects are obtained on the Sykes sharpened tools since each lip on the cutting edges is formed individually and therefore positioning errors between

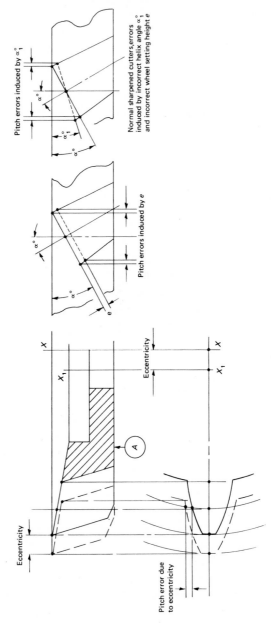

Fig. 189. Normal sharpened cutter's errors induced by incorrect sharpening

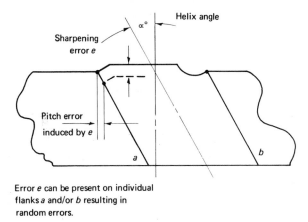

Fig. 190. Sykes sharpened cutters – errors induced by incorrect sharpening

the wheel and the tooth give the condition shown in Fig. 190. As each cutting edge is sharpened individually the error induced by faulty wheel positioning can result in random errors and great care must be exercized to maintain a common plane for each flank *a* and *b*.

2.5.2 Hobs

(a) *Methods of sharpening*

Prior to the advent of the modern automatic hob sharpening machines this type of tool was ground on a tool and cutter grinder with a special guide. This guide unit was made with a series of notches equivalent to the number of teeth in the hob and with the same direction and lead as the helix angle. The guide unit was placed on the same arbor as the hob and the whole assembly mounted between centres on the cutter grinder table. A finger on the wheel head engaged in the notches on the guide unit so that as the table was reciprocated the stationary finger caused the hob to follow a helical path across the face of the wheel. In order to allow the wheel to run out of the hob flute for indexing the guide was made longer than the hob so that the finger did not lose contact. To index from one flute to another the finger was usually spring-loaded so that it could be deflected by hand and the guide unit rotated on its axis to the next tooth. The accuracy produced on the hob using this technique was dependent upon the accuracy of the guide unit, which was usually supplied by the hob manufacturer. Fig. 191 shows the system

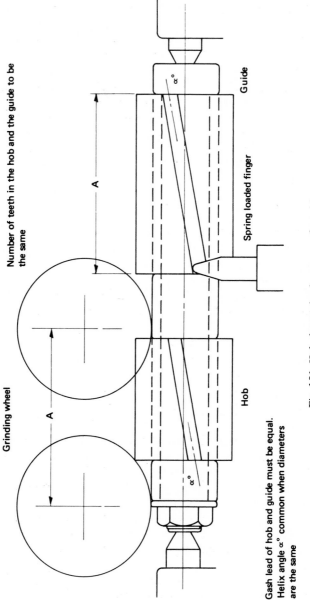

Number of teeth in the hob and the guide to be the same

Grinding wheel

Spring loaded finger

Guide

Hob

Gash lead of hob and guide must be equal.
Helix angle α° common when diameters
are the same

Fig. 191. Hob sharpening by means of a guide

in kinematic form; its advantage is that special sharpening machines are not necessary, and it is ideal if only small quantities of hobs have to be sharpened. For hob sharpening in larger quantities, however, it is necessary to consider the special machines designed specifically for the purpose. The desirable features of a hob sharpening machine could be summarized as follows:

(a)   On machines with capacity up to 250 to 300mm diameter the hob should remain stationary while being ground, the only movement being for indexing or rotation around its own axis for following the helical path of the flute. All axial reciprocation should be performed by the grinding wheel head since the mass being moved is constant and this means the inertia effects encountered when reversing can be taken into account in the design. If the hob is reciprocated the mass is not constant since the weight of the hob alone can vary from a few pounds up to 280 lb in the case of a 250mm x 300mm hob. The vibration and shock loads from out of balance forces can cause inaccuracies.

(b)   Large diameter tapered grinding wheels are preferable in order to keep wheel wear to a minimum.

(c)   Accurate positioning of the wheel relative to the centre line of the hob is essential and facility to raise or lower the wheel relative to the flute face is desirable.

(d)   Indexing accuracy is of prime importance and large diameter index plates mounted on taper bores to eliminate eccentricity are desirable features.

(e)   Axial positioning of the work to the wheel independent of the stroke is necessary in order to prevent excessive overtravel.

(f)   Some form of anti-backlash is necessary in the drive to the helical motion in order to prevent any play between the wheel and the flute face at the reversal of the stroke.

(g)   Some hobs are integral with a shaft (see Fig. 192) so that the wheel cannot pass clear through the flute for indexing. Some facility must be provided therefore for an automatic cycle whereby the wheel (or the hob) can be displaced radially and returned to the start position prior to indexing and commencement of the next cycle.

(h)   Some form of automatic cycle is necessary and should include (i) automatic wheel dressing after a predetermined number of strokes; (ii) incremental feed after a predetermined number of indexing, plus control of the total feed movement for stock removal; and (iii) control of wheel speed for grinding and dressing.

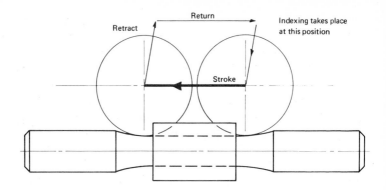

Fig. 192. Automatic hob sharpening; no grinding on return stroke owing to lack of wheel run-out for indexing

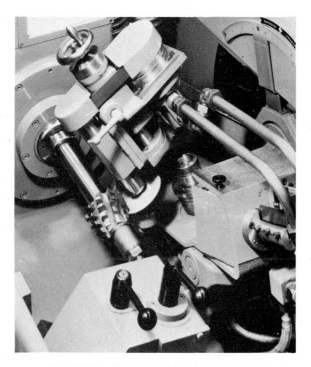

Fig. 193. Generating wheel dressing attachment

(j)   Generating wheel dressing attachments are necessary to maintain radial flute faces when producing hobs of high helix angle, as
shown in Fig. 193.

(b) *Effect of errors*

The extent to which the quality of hobbed teeth depends upon the
use of correctly sharpened hobs can be appreciated when the manner in
which the hob generates the tooth flanks is considered. We have seen in
previous chapters that a series of enveloping cuts are taken by the
cutting edges of the hob and provided that these generating cuts are
taken at the correct intervals perfect results can be obtained. If the hob
is incorrectly sharpened, however, the cutting edges deviate from the
theoretical position in respect of both shape and position and malform
the gear even though other elements of the hob may be accurate.

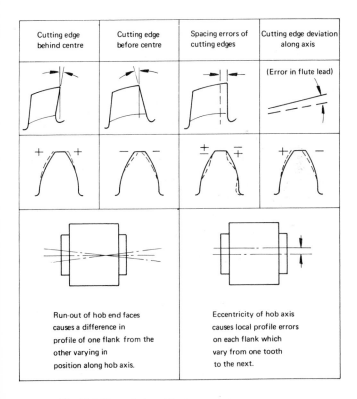

Fig. 194. Errors induced by incorrect hob sharpening

The principal elements to be checked and maintained when sharpe
ing are (a) the radiality of the flute face; (b) the adjacent pitch error
the flutes; (c) the cumulative pitch error of the flutes, and (d) t
deviation from the theoretical path of the flute lead or helix ang
Errors in any one or all of these elements can cause errors of the natu
shown in Fig. 194.

It is extremely important that the hob be mounted true both wh
sharpening and when in use on the hobbing machine. Run-out of t
hob axially is just as important as radial run-out.

PART 2.6: MATERIAL AND HEAT TREATMENT

Most cutting tools are made in one of the high-speed steels, so called because of their ability to maintain a cutting edge at high machining rates even up to the point where the tool is red hot. Forging or rolled bar stock is used and a considerable difference of opinion exists as to which gives the best results, except on tools over 125 mm diameter which are almost certainly manufactured from individually forged blanks.

Fig. 195b shows a range of steels which have been standardized by the American Iron & Steel Institute (A.I.S.I.) and gives the main constituents and the effect they have on abrasion resistance, toughness, red hardness, etc. The M2 steel is very popular in the U.S.A. and in the U.K. and is used extensively in the manufacture of accurate unground tools. Fig. 195a shows roughly the heat treatment applied to high-speed steels but it should be appreciated that this is only a guide and nothing can replace practical experience in this respect. The hardness of the tool can range between 62 and 67 C Rockwell according to the type of steel, but the most usual figure is 63 to 64 C Rockwell. Fig. 196 shows the heat-treatment plant at W. E. Sykes which specializes in high speed steel and is typical of a modern installation. After heat treatment the performance of a tool can sometimes be improved by some form of secondary treatment providing the temperatures required are low enough to avoid annealing of the cutting edges or distortion of the

High speed steel heat treatment process chart

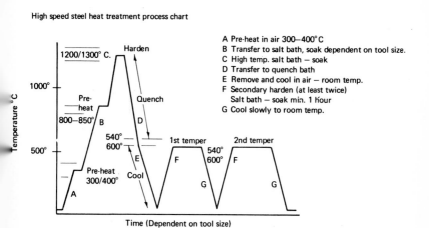

A Pre-heat in air 300–400°C
B Transfer to salt bath, soak dependent on tool size.
C High temp. salt bath – soak
D Transfer to quench bath
E Remove and cool in air – room temp.
F Secondary harden (at least twice)
   Salt bath – soak min. 1 hour
G Cool slowly to room temp.

Time (Dependent on tool size)

Fig. 195a. Typical heat treatment times

Type	Tungsten	Molyb.	Chrome	Vanadium	Cobalt	Edge Toughness	Red Hardness	Abrasive Resistance	Grindability	Application
T1	18		4	1						General purpose
T2	18		4	2						Light cuts at high speed
T3	18		4	3						High red hardness
T4	18		4	1	5					Ditto and abrasion resistant
T5	18		4	2	8					Heavy stock removal – hard materials
T6	20		4.5	1.5	12					Heavy planer tools
T7	14		4	2						Roughing hard materials
T8	14		4	2	5					Light cuts at high speed
T9	18		4	4						Good red hardness – abrasion resistant
T15	12.5		4.75	5	5					General purpose
M1	1.5	8.5	4	1						
M2	6	5	4	2						Fine edge tools
M3	6	6	4	2.4/3						Abrasion resistant
M4	5.5	4.5	4.5	4						Heavy cuts
M6	4	5	4	1.5	12					Fine edge tools
M7	1.75	8.75	4	2						
M8	5.5	4.5	4.25	1.5						General purpose
M10		8	4	2						
M15	6.5	3	4.75	5	5					Good red hardness – abrasion resistant
M30	1.5	8.25	4	1.25	5					" "
M34	1.75	8.75	4	2	8.25					" "
M35	6	5	4	2	5					Heavy cuts at high friction
M36	6	5	4	2	8					
M38	6.5	5	4	2	9					
M40		8	4	1.5	8					

High speed steels A.I.S.I. Series

Fig. 195b. Heat treatment applied to high speed steels

Fig. 196. Heat treatment plant at W. E. Sykes

profiles. The spraying of hard metals on to the finished tooth profiles is not usually satisfactory owing to the difficulty of maintaining an even coating, particularly on the cutting edge itself. Coatings can be applied to the flute face or the tooth flanks but where the two areas join to form the cutting edge it is virtually impossible to maintain an edge. For this reason they are seldom of any practical use and better results are obtained by creating a hard skin or layer on the parent metal itself. Probably the best technique is nitriding where a surface hardness up to some 70 C Rockwell can be obtained, the depth of case depending on the length of time in the bath. The finished tools are immersed in a cyanide bath at about 550°C for periods of one hour which should give a case approximately 25 microns deep. The steels respond according to their ability to absorb nitrogen and in this respect the molybdenum types are possibly better than the tungsten ones.

Vapour blasting is popular and often used by tool manufacturers as a standard treatment for tools before despatch to the customer. The tool is not altered in its structure metallurgically since no heat treatment is involved, but the finished surfaces are blasted by a jet of water containing fine abrasive particles. This produces a matt finish and has the effect of honing by removing the fine grinding burrs which are usually present particularly on the cutting edges. Great care must be exercised to see that the treatment is not excessive since this can also dull the edges by removing metal as distinct from grinding rags from the corners. Experiments are still going on with oxide coatings and steam treatments and tools have been given a form of stress relieving by ultrasonic vibrations but insufficient data are available to form any conclusions regarding their effectiveness.

SECTION 3

INSPECTION

In view of the stringent requirements of modern gear production organizations it is essential that both the machine and cutting tool be of a high order of accuracy. This requirement has in turn led to the development of a range of sophisticated machines for checking both tool and machine.

PART 3.1: CHECKING THE CUTTING TOOL

3.1.1 Hobs

Fig. 197 shows the current BS 2062 for gear hobs for general purposes. It can be seen that it covers a number of grades of accuracy – grades *AA, A* and *B* are for ground-form hobs and grade *D* for unground hobs. In addition to indicating the standard of accuracy it shows the recommended method for checking each element.

The checks on the first three elements require no explanation or special equipment since the hob can be mounted between a pair of bench centres and checked for accuracy. Elements 4 to 6, however, require special checking equipment and it is desirable that permanent records of accuracy be taken, so some form of electronic recorder should be used.

Fig. 198 shows the developed path of the true helix of the hob and serves to indicate the difference between elements 4 and 5. It can be seen that without element 5 it would be possible to observe element 4 but still have a cyclic error per revolution of the hob.

The lead checking machine required for checking this element is made specially for hobs and worms and is not suitable for checking the leads of gears, as it would not be possible to cover the range. One of the best checking instruments for this type of work is the Klingelnburg hob tester (Fig. 199). This machine is fully universal in that it will check all the important elements of a hob – lead, base pitch, pressure angle, base helix angle, flute pitch, thread spacing and so on – without removing the tool from its datum position. The machine consists essentially of three main elements: the base carrying the longitudinally travelling rolling slide, the vertical column with work drive, and the column with the measuring head containing the electronic stylus and recording unit. The hob is mounted between centres and is driven by two friction rolling discs on a common axis. The rotary motion of the friction discs and workpiece is obtained by contact with a guide bar on the longitudinal slide. The rolling slide also carries a sine bar which can accurately be set to the required angle and which permits vertical

PERMISSIBLE ERRORS—GRADES AA, A, B AND D HOBS
(Expressed in ten-thousandths of an inch)

Element	No.	Illustration	Test	Grade	Preferred diametral pitch (see Note 2 for non-preferred)								
---------	-----	--------------	------	-------	1, 1¼, 1½	2 and 2½	3	4	5	6, 7, 8	10 and 12	16	20
Hub reference Datum surfaces	1	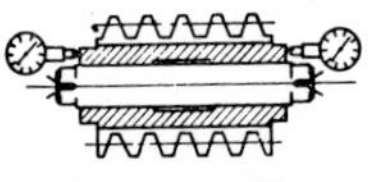	Axial run-out	AA	—	3	2	2	2	2	2	2	2
				A	8	5	2	2	2	2	2	2	2
				B	10	8	4	4	3	3	2	2	2
				D	10	8	5	5	4	4	3	3	3
	2	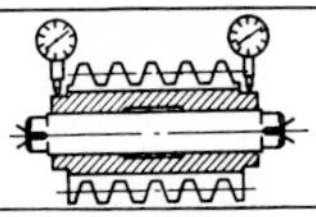	Radial run-out	AA	—	4	3	3	3	3	2	2	2
				A	10	5	4	3	3	3	2	2	2
				B	12	8	6	5	4	4	3	2	2
				D	15	10	8	8	6	6	5	5	5
Outside diameter	3	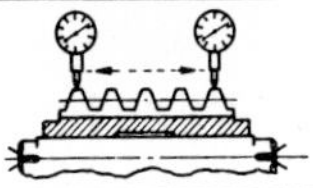	Radial run-out	AA	—	15	10	10	10	10	10	5	5
				A	30	20	15	15	10	10	10	10	10
				B	40	30	25	20	15	15	15	10	10
				D	60	55	50	45	35	35	30	25	25
Tooth lead variation	4	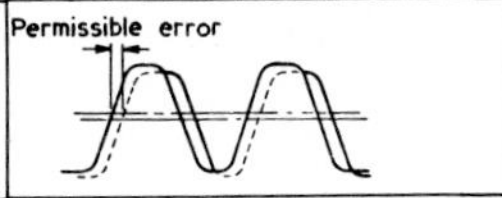 Permissible error	Variation along tooth helix from tooth to tooth	AA	—	3	2	2	2	2	1·5	1·5	1·5
				A	7	5	4	3	2	2	2	2	2
				B	10	8	6	4	3	3	3	3	3
				D	25	20	16	14	12	10	10	8	8
	5	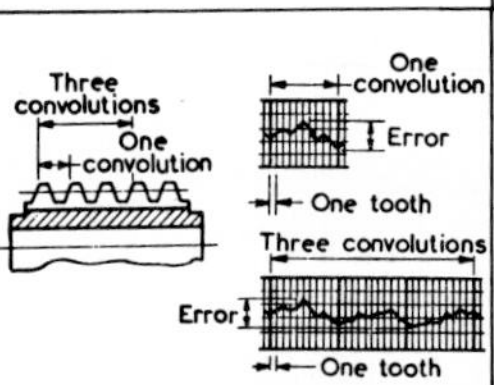 Three convolutions One convolution One convolution Error One tooth	Variation along tooth helix in one convolution (cyclic pitch error)	AA	—	6	3·5	3·5	3	3	3	2	2
				A	25	18	10	8	6	5	5	4	4
				B	35	25	17	11	9	7	7	6	6
				D	60	50	40	30	25	20	20	18	18
	6	Three convolutions Error One tooth	Variation along tooth helix in any three consecutive convolutions (cumulative pitch error)	AA	—	12	6	5	4	4	3	3	3
				A	38	26	15	12	9	8	8	7	7
				B	53	38	22	16	12	11	10	9	9
				D	120	100	80	60	50	40	35	25	25
Tooth	7	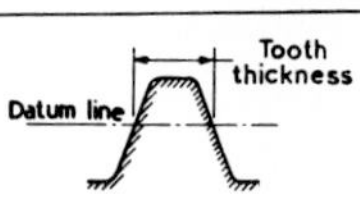 Tooth thickness Datum line	Thickness (minus only)	AA	—	15	10	10	10	10	5	5	5
				A	30	20	15	15	10	10	10	10	10
				B	30	20	15	15	10	10	10	10	10
				D	40	35	30	25	20	20	20	20	20
	8	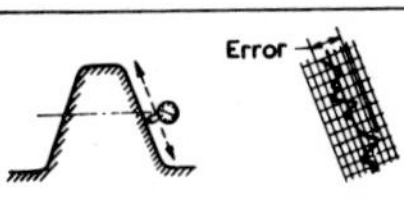 Error	Permissible error over the straight flank from the design form	AA	—	4	3	2	2	2	1·5	1·5	1
				A	10	5	3	3	2	2	2	2	2
				B	16	8	5	5	4	3	3	3	3
				D	80	55	30	18	12	8	8	6	6

Fig. 197. Accuracy chart for gear hobs

ovement of the measuring head during the roll movement. The stylus
nit is mounted on the upper surface of the measuring head and can be
djusted relatively to the hob.

A trace of the cutting edges can be taken thus showing the total
rrors in the hob — the stylus unit is set over to the base circle and the
heck taken along the line t_e as shown in Fig. 200.

Deviations of the cutting faces from their theoretical shape and
osition are recorded by the tracer head for each tooth. The result is a

PERMISSIBLE ERRORS—GRADES AA, A, B AND D HOBS
(Expressed in ten-thousandths of an inch)

Element	No.	Illustration	Test	Grade	1, 1¼, 1½	2 and 2½	3	4	5	6, 7, 8	10 and 12	16	20
					Preferred diametral pitch (see Note 2 for non-preferred)								
Tooth (continued)	9		Start of profile modification (plus or minus)	AA	—	140	120	100	80	60	40	30	20
				A	200	180	160	140	120	100	80	60	40
				B	220	200	180	160	140	120	100	80	50
				D	260	240	220	200	180	160	140	120	120
	10		Symmetry of start of profile modification	AA	—	80	80	60	50	40	30	15	15
				A	150	130	120	100	90	80	60	50	35
				B	180	150	130	120	100	90	80	70	45
				D	200	180	160	140	120	110	100	90	90
Cutting faces of gashes	11		Adjacent spacing error	AA	—	20	10	10	10	10	10	10	10
				A	40	30	25	20	15	10	10	10	10
				B	50	45	40	30	20	15	15	10	10
				D	60	60	50	50	30	25	25	20	20
	12		Cumulative spacing error	AA	—	40	30	30	30	30	20	20	15
				A	80	60	50	40	30	30	30	25	25
				B	100	90	80	60	50	50	50	40	35
				D	120	120	100	100	80	80	70	60	60
	13		Radial alignment of gash over cutting depth	AA	—	10	10	6	6	5	5	5	5
				A	30	15	10	8	6	5	5	5	5
				B	50	25	15	10	8	7	7	5	5
				D	100	75	50	40	30	20	20	15	15

	14		Departure (plus or minus) from angle of gash (expressed as a linear dimension of ten-thousandths per inch of face width)	Grade	AA	A	B	D
					6	10	10	15

	15		Diameter (plus only)		2 and 2½	1½	1¼	¾
Bore				AA	5	3	3	2
				A	8	5	3	2
		Standard bores shall be finished straight and parallel, within the tolerances given, for 75 per cent of each bearing length for Grade AA, A and B hobs and 50 per cent for Grade D hobs.		B	10	8	4	3
				D	10	8	5	4

Tooth relief	16	These tolerances apply to new and resharpened hobs		AA A B D	The accuracy of the flank relief and top relief shall be such that the maximum permissible errors specified will not be exceeded when the cutting faces are reground to the above tolerances.

Fig. 197 *(contd.)* Accuracy chart for gear hobs

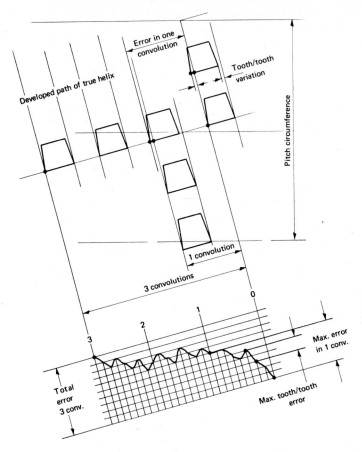

Fig. 198. Developed path of true helix of hob

chart showing the cumulative errors resulting from flank errors on all cutting edges over the length of the hob, lead errors, diameter runout and cyclic errors.

Checks 8, 9 and 10 can also be taken on this instrument, the stylus moving along the line *AC* in the diagram and the error automatically being recorded on a chart.

Helical surfaces produced by a straight-sided generator can be checked in this manner — involute helicoids or helicoids with straight sides in either the axial or the normal planes. A simpler machine for

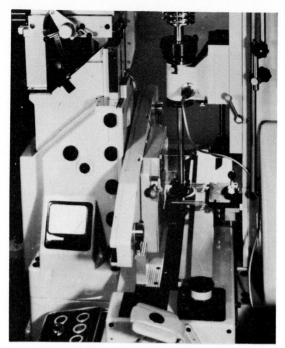

Fig. 199. Klingelnburg hob testing machine

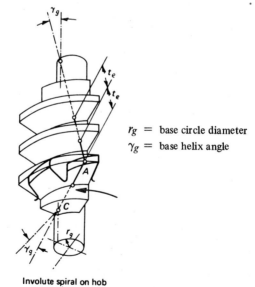

r_g = base circle diameter
γ_g = base helix angle

Involute spiral on hob

Fig. 200. Accuracy check over cutting edges of a hob

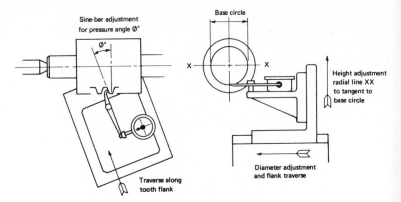

Fig. 201. Flank angle checking of hobs and worms

performing the same function is shown in Fig. 201. The compound slide can be set to the required angle, i.e. the base helix angle with the stylus set below the centre line and tangential to the base circle or with the stylus on the centre line and set to the pressure angle in the axial or normal planes. The stylus is traversed, therefore, in the plane containing the straight line generator and any angle error is indicated as a linear measurement on the dial indicator.

When checking hobs with helical flutes the cutting edges lie in a plane normal to the helix and the profile of the thread in the axial plane is affected by the variation of the lead angle brought about by the relief on the tooth. The flank angle of the leading edge is therefore greater and the trailing edge smaller in the axial plane as shown in Fig. 202.

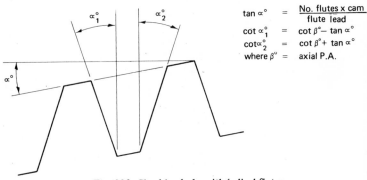

$$\tan \alpha° = \frac{\text{No. flutes} \times \text{cam}}{\text{flute lead}}$$

$$\cot \alpha_1° = \cot \beta° - \tan \alpha°$$

$$\cot \alpha_2° = \cot \beta° + \tan \alpha°$$

where $\beta°$ = axial P.A.

Fig. 202. Checking hobs with helical flutes

Fig. 203. Klingelnburg flute checking machine

The elements 11, 12 and 13 are the adjacent, cumulative and radial flute errors and these can all be checked on one simple instrument such as a pair of bench centres and a divide plate. Alternatively a more sophisticated machine for checking the same elements plus the flute lead (element 14) is the Klingelnburg PWB. 300 shown in Fig. 203.

The measuring slide is adjustable axially and contains two dial indicators which can also be adjusted radially inwards. The radiality of the flute face is checked by setting a dial gauge to zero (relative to the hob centre line) by means of a gauge and then moving it radially in over the face of the flute and reading off the deviation. Spacing is checked both by using dial indicators and by comparing the individual readings against the first reading — cumulative errors are obtained by plotting the adjacent errors graphically. Checking the flute lead is achieved by taking a series of readings along the flute face. The dial indicator carriage is moved precisely along the hob axis by means of slip gauges and the necessary radial adjustment to the hob is made by means of a sine bar arrangement.

3.1.2 Shaper cutters

BS 2887 covers spur type tools only, from 4 to 20 D.P. Helical cutters are not yet covered by a British Standard. Fig. 204 shows the elements to be checked and the tolerances allowed for the two grades of accuracy *AA* and *A*. The elements 1 to 7 inclusive are realtively simple to check and require no specialized equipment. The element 8 is involute profile and is checked on a special machine which is described in detail on page 290. The permissible error on profile must be plus at both tip and root in order to ensure that the profile deviation produced on the gear is negative and provides a certain degree of relief on the flanks. When checking the profile on the cutting edge a broad-faced stylus is used so that it is enveloped and the pressure angle is then the standard value. For checking behind the cutting edge, however, in the transverse plane, a knife edge stylus is used and the pressure angle deviates from the standard angle owing to the front face and top relief angles. The angle can be found from the relation

$$\tan^{-1}\left(\frac{\tan 20^\circ}{1-\tan 5^\circ \tan \alpha_t^\circ}\right),$$

where α_t° = top angle; and the front face angle is 5°.

TOLERANCES
Units 0·0001 in. unless otherwise stated

Element No.	Typical test arrangement	Element	Test	Nominal cutter size	Grade of cutter	Permissible error				
						Diametral pitch				
						4 to 4·999	5 to 5·999	6 to 8·999	9 to 12·999	13 to 20
1	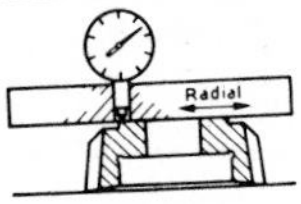	Datum face (radial) (see Note 2)	Flatness (concave only) of datum face	Up to 3 in.	AA	1	1	1	1	1
					A	2	2	2	2	2
				4 in.	AA	1	1	1	1	1
					A	2	2	2	2	2
2	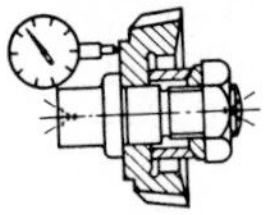	Datum face (See Note 2)	Axial run out	Up to 3 in.	AA	1	1	1	1	1
					A	2	2	2	2	2
				4 in.	AA	2	2	2	2	2
					A	3	3	3	3	3
3	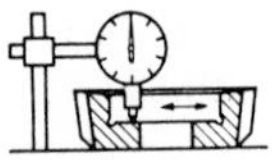	Inner face (See Note 2)	Parallelism of inner face with datum face	All sizes	AA	1	1	1	1	1
					A	2	2	2	2	2
4	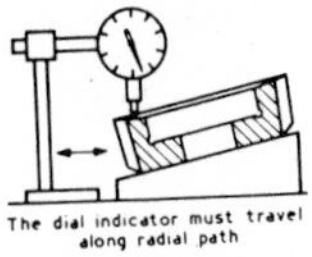	Cutting face (sharpened)	Accuracy of cutting rake (5° nominal)	All sizes	AA & A	+ 0' −15'	+ 0' −15'	+ 0' −15'	+ 0' −15'	+ 0' −15'
5	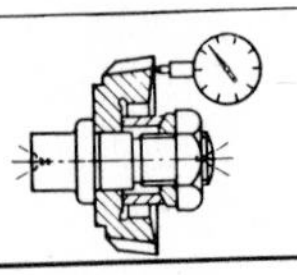	Cutting face (sharpened)	Axial run-out	Up to 3 in.	AA & A	3	3	3	3	3
				4 in.	AA & A	4	4	4	4	4

The dial indicator must travel along radial path

Fig. 204. Tolerances for gear shaper cutters

When checking helical cutters the profile should be checked using a broad stylus so that it envelops the cutting edge. Although cutters with Sykes type sharpening can be checked on, or behind, the cutting edge (there is no front rake) the tools with normal sharpening should always be checked on the cutting edge.

Element 9 is a comparative check requiring no special equipment and the error recorded is a composite of several elements, the principal one being the concentricity of the teeth when being ground.

Element 10 covers both adjacent and cumulative pitch errors and these are extremely important factors in the gear shaping process. The adjacent pitch error is relatively simple to measure and does not necessarily require a special machine but the cumulative test requires

TOLERANCES (continued)
Units 0·0001 in. unless otherwise stated

Element No.	Typical test arrangement	Element	Test	Nominal cutter size	Grade of cutter	Permissible error Diametral pitch				
						4 to 4·999	5 to 5·999	6 to 8·999	9 to 12·999	13 to 20
6	Datum cylinder	Outside diameter	Radial run-out, relative to bore	All sizes	AA	3	3	3	3	3
					A	5	5	5	5	5
7		Datum cylinder (see above) (See Note 3)	Run-out relative to bore	All sizes	AA	1	1	1	1	1
					A	2	2	2 Where applicable	2	2
8	Design profile is represented by full line.	Tooth (See notes 5 & 6)	Profile error. Excess metal only at tip and cutting depth	All sizes	AA	2·5	2	2	1	1
					A	3	2·5	2	1·5	1
9	Total range of indicator reading measured over a ball contacting near the reference circle in each tooth space.	Tooth	Run-out of reference circle, relative to bore	All sizes	AA	2	2	2	2	2
					A	5	5	5	5	4
10		Tooth	Adjacent pitch	All sizes	AA	2	2	2	1	1
					A	3	3	3	2	2
		Tooth	Cumulative pitch (over approx. 180°)	Up to 3 in.	AA	3	3	3	3	3
					A	6	5	5	5	4
				4 in.	AA	4	4	4	4	4
					A	8	7	6	5	5

Fig. 204 *(contd.)* Tolerances for gear shaper cutters

some consideration. Since the adjacent error can easily be measured the cumulative is often arrived at by graphical analysis of the adjacent errors — the errors are plotted and the cumulative deduced from them. When dealing with extremely fine measurement, however, this method leaves something to be desired. Consider the following example.

(a)   instrument for measuring adjacent pitch error accurate within 1 micron

(b)   100 teeth to be measured.

It is possible that although the measurement of the adjacent pitch is extremely accurate the measuring error will accumulate when plotting the results over 100 teeth — this could therefore be 100 microns. This could be more than the tolerance allowed on the element being measured, and even assuming that the errors when plotted only accumulated by 10%, there would still be an inaccuracy of 10 microns.

The measurement of cumulative errors should therefore be direct since then a tooth could be measured at 180° to the datum and the error in measurement would only be the error from the means of recording it. As the error in the recording method is known it is possible to establish the cumulative error more accurately than with the graphical technique. The direct recording of cumulative errors is usually made by angular measurement and the accuracy of the machine can be determined by a master polygon.

3.1.3 Rack tools

These are covered by BS 2697 and some of the elements to be checked and the tolerances allowed are shown in Fig. 205. The checking of the pitch is a straightforward linear measurement but again a direct measurement of the cumulative error is to be preferred. The profile checks for symmetry of form, profile deviation and angle error can be performed optically and require no special explanation.

3.1.4 Shaving tools

These are covered by BS 2007 (1959) and the principal elements for checking are clearly shown in the standard. The profile, lead and pitch errors would be checked in the same manner as already described for gear shaper cutters, although the lead check would show an interrupted trace due to the slots down the flanks. The slots themselves are usually provided down the tooth flanks in annular form so that they are in-line from one tooth to the next. Some manufacturers claim, however, an

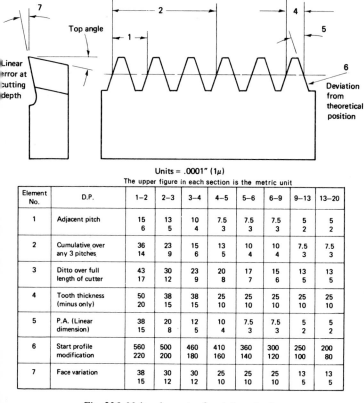

Units = .0001" (1μ)
The upper figure in each section is the metric unit

Element No.	D.P.	1–2	2–3	3–4	4–5	5–6	6–9	9–13	13–20
1	Adjacent pitch	15	13	10	7.5	7.5	7.5	5	5
		6	5	4	3	3	3	2	2
2	Cumulative over any 3 pitches	36	23	15	13	10	10	7.5	7.5
		14	9	6	5	4	4	3	3
3	Ditto over full length of cutter	43	30	23	20	17	15	13	13
		17	12	9	8	7	6	5	5
4	Tooth thickness (minus only)	50	38	38	25	25	25	25	25
		20	15	15	10	10	10	10	10
5	P.A. (Linear dimension)	38	20	12	10	7.5	7.5	5	5
		15	8	5	4	3	3	2	2
6	Start profile modification	560	500	460	410	360	300	250	200
		220	200	180	160	140	120	100	80
7	Face variation	38	30	30	25	25	25	13	13
		15	12	12	10	10	10	5	5

Fig. 205. Major elements of rack type tools

mprovement in shaving conditions by providing the slots in a differ-
ntial pattern so that successive teeth are not in line and under these
ircumstances the pitch of the slots relative to each other must be
hecked. This should not be confused with the requirement for stagger-
ng the slots due to the tools being used for underpass shaving. In this
atter case the slots must be displaced to each other in order to cut and
he degree of displacement (as long as it is not a multiple of the pitch)
s not important. Where the slots are displaced to each other great care
nust be taken to see that where the slots break out into the side faces
he teeth partly formed are removed completely.

PART 3.2: CHECKING THE GEAR

Dealt with in detail this could be the subject of a separate book, as there are now a vast number of machines and techniques available for checking the various elements. It is proposed, therefore, to deal only with the more important elements and the most widely used techniques.

Fig. 206 summarizes the situation. It can be seen that there are two main groups: (a) elemental checks and (b) composite checks. As the name suggests the elemental checks give a direct measurement of individual errors such as profile, lead, pitch etc., whereas the composite

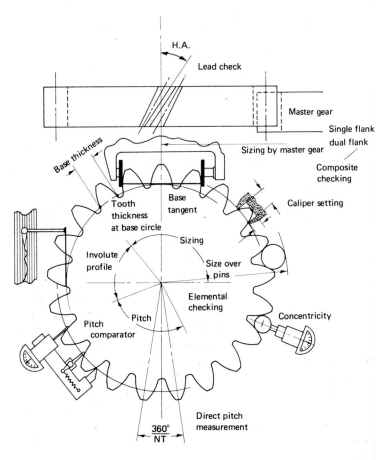

Fig. 206. Methods of checking gear elements

heck contains the errors from several elements and it is necessary to
deduce the error in any single element by analysis.

Involute	(a)	elemental —	involute checking machines
	(b)	composite —	single flank checking with master gear
Lead	(a)	elemental —	lead checking machines
	(b)	composite —	master gear, visual reading
Pitch	(a)	elemental —	pitch comparator, optical table
	(b)	composite —	single flank test with master
Concentricity	(a)	elemental —	ball check in tooth space
	(b)	composite —	single and dual flank test
Size	(a)	elemental —	comparator — base tangent — caliper
	(b)	composite —	master gear, dual flank test.

The tolerances for the above elements are covered by a number of
gear standards, the most common in Europe being probably the BS 436,
the German DIN Standard 3963 and the Fine Pitch Standard BS 978,
while in the U.S.A. the AGMA standard is the most prevalent.

3.2.1 Involute checking machines

The basic principle of the involute checking machine is very
simple as can be seen from Fig. 207. A disc equal to the base circle

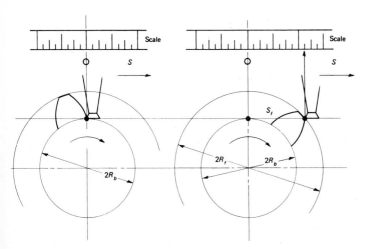

Fig. 207. Basic principle of the involute tester

diameter of the gear to be checked is in contact with the cross slide S so that the friction between the surfaces causes the disc to rotate around its own axis; the slide S moves in the direction of the arrows. The stylus is set so that its tracing point is on a line tangential to the base cylinder and remains in contact with the tooth flank as the cylinder or disc is rotated. The locus of the tracing point is therefore an involute and any deviation of the tooth flank from a true involute path is picked up by the stylus and recorded on an indicator or chart. The amplitude of the error is shown on the chart but in order to determine the location of the error the slide S is graduated so that the distance along the line of action can be measured.

From the diagram it can be seen that

$$R_f = \sqrt{S_f{}^2 + R_b{}^2}$$

Although separate lead and involute checking machines are available the two functions are combined on most modern checking machines.

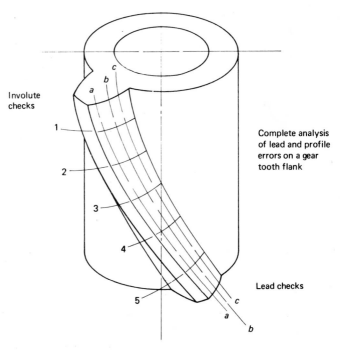

Fig. 208. Complete analysis of lead and profile errors on a gear tooth flank

These combined machines permit the checking of the involute and lead without changing the setting of the machine or the work. Although it is usual to check a tooth at one position only the combination of the two tests has the advantage that is possible to envelop the tooth flank by a series of readings thus charting the whole tooth if necessary. Fig. 208 shows the principle; here 1 - 5 are profile lines and a - c are lead lines, but any number of readings can be taken.

3.2.2 Checking the helix

On the machines equipped for checking both lead and involute profiles the basic kinematics of operation are shown in Fig. 209. When

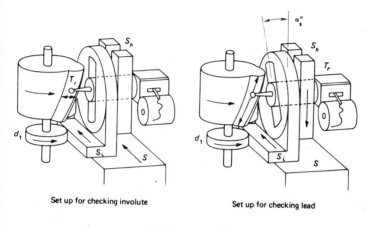

Set up for checking involute   Set up for checking lead

Fig. 209. Combined lead and involute checkers

checking the involute the compound slides S and S_i are locked together and the vertical slide S_h and sine bar are disconnected. The cross slide S_i is traversed as indicated by the arrow and this imparts a rotation to the base circle disc d_1, whereas the tracer head T_r does not move relative to the slides S - S_i and S_h but constitutes the straight line generator relative to the gear axis. When checking the lead or helix angle the slide S is disconnected from S_i and the slide S_h and the sine bar is engaged. As the tracer head T_r is traversed vertically along the slide S_h motion is imparted to the slide S_i by the inclination of the sine bar a. The resultant motion of the tracer or stylus relative to the gear flank is a helical path which corresponds to the theoretical helix angle.

If the rolling disc d_1 is equal to the base circle then the sine bar setting is equal to the base helix angle. For other diameter discs the sine bar setting $\alpha_s{}^\circ$ can be found from the relation

$$\tan \alpha_s{}^\circ = \tan \alpha_b{}^\circ \; \frac{d_1}{d_g} \quad,$$

where $\alpha_b{}^\circ$ = base helix angle
d_g = base circle

Typical of the combined lead and involute checking machines is the Klingelnburg PFS. 600 shown in Fig. 210.

Lead testers

Before discussing the method of testing it is worth considering the type of component to be checked since the requirements for checking

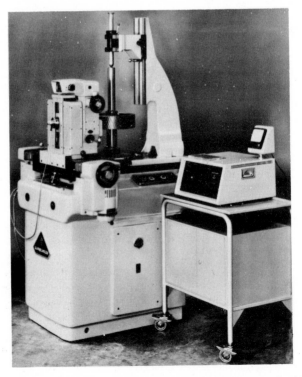

Fig. 210. Klingelnburg PFS.600 machine for lead and involute checking

a helical gear of say 76.2mm P.C.D. 10 N.D.P. 30° H.A. are entirely different from those required for a single start worm 10 N.D.P. 76.2 P.C.D. The helical gear would have a lead of

$$\frac{76 \cdot 2 \times \pi}{\tan 30°} = 414.635 \text{ mm},$$

whereas the worm would have a lead of $\frac{7.9796}{\cos \alpha}$.

where $\sin \alpha = \dfrac{\pi}{10 \times \pi \times 76 \cdot 2}$

$= 0.0333,$

lead $= 7.984$ mm.

For helical gears, therefore, the lead requirements are likely to be of the order of 150mm to infinity whereas with hobs or worms zero to 50mm covers most conditions.

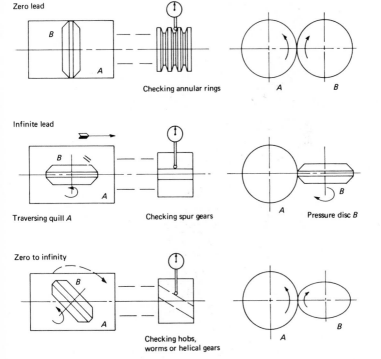

Fig. 211. Principle of the Goulder lead tester

Most lead checking machines operate on a sine bar or drum and band principle and they are therefore limited to checking either helical gears or worms but not both. The sine bar is not efficient over 45° and at 85° it is not possible to translate the motion from one plane to another. Only one range of leads can therefore be covered and this dictates the plane in which the sine bar operates. The drum and band principle is not suitable for short leads since it entails several revolutions of the drum, which are not possible. Rack and pinion movements partly overcome the problem but an excessive length of rack is then entailed to give the required translation of movement. For this reason most lead checking machines are limited to checking a certain class of work and it

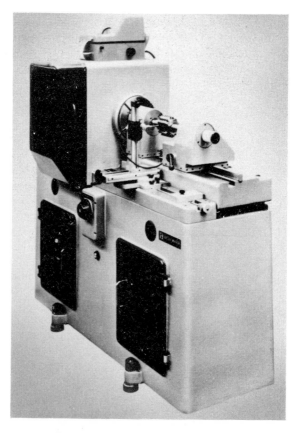

Fig. 212. Goulder machine checking a hob

is therefore worth making special mention of the Goulder lead tester which overcomes these basic problems by a unique method enabling leads from zero to infinity to be checked.

Goulder lead tester

The basic principle of the machine is shown in Fig. 211. A pressure disc B can be preloaded against the main spindle A. If the axes of A and B are parallel to each other then when the disc B rotates so also does A and there is no movement of A axially resulting in zero lead. When the axes of A and B are at $90°$ to each other, for a given rotation of B the spindle A does not rotate but moves along its own axis at a rate determined by the peripheral speed of B. The result is therefore an infinite lead which enables spur gears to be checked. If the axes of A and B are inclined to each other then helical gears can be checked since for a given rotation of B the spindle A rotates and moves along its own axis, the rates of motion being dependent on the inclination of the two axes. Using this unique principle, therefore, the machine has the ability to check spur gears, annular rings, helical gears of all angles and hobs or worms.

The machine is illustrated in Fig. 212 where it is checking a hob. Here the means of setting the inclination of the axes is by an optical scale. The pressure disc B is powered by a motor and can be disengaged from the main spindle when required. Deviations from the true path on the work are picked up by means of a suitable stylus and recorded on an electronic recorder.

3.2.3 Pitch checking

A machine built specially for the automatic checking and plotting of pitch errors is the Coventry Gauge & Tool Matrix. The basic principle is shown in Fig. 213. The drive piston actuates the sine bar which locks on to the drive ring and rotates the faceplate until the sine bar is arrested by the fixed stop. The drive ring is held by a further clamp while the sine bar disengages and returns to its start position. While the drive ring is clamped the probe unit advances automatically, probes the gear flank and then retracts. This starts the cycle again until all the readings have been taken. The machine, complete with graphical recorder and sensing head, is shown in Fig. 214.

A method used by W. E. Sykes and favoured by the author involves direct measurement of both adjacent and cumulative pitch error from a

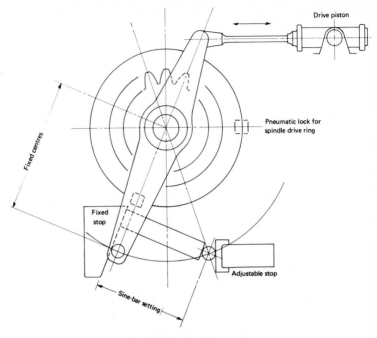

Fig. 213. Automatic pitch checking

precision optical table. The table itself is built to a high degree of precision (2 seconds of arc) and is then fitted with a tape control system which gives an automatic programmed cycle and more important enables the known errors in the table to be allowed for by interpolation. The table as shown can therefore be controlled automatically within 1 second of arc and the errors are recorded by an air operated sensing head.

There are a variety of attachments available for checking the tooth pitch spacing by comparative methods. These all operate on virtually the same principle as in Fig. 215. The spring-loaded feeler rotates the gear so that it abuts the fixed stop while the measuring feeler registers the position of a tooth flank on a dial indicator or sensing head. For accurate results the preload of the gear against the fixed stop and the preload of the measuring feeler against the flank being recorded must be the same for all positions and this is achieved by means of the tension springs.

Fig. 214. Coventry Gauge & Tool Matrix for automatic pitch checking

3.2.4 Concentricity

This is difficult to measure as an element on its own as it is virtually impossible to divorce it from the other errors — pitch error, local form errors and so on. The accepted method is to place a ball in the tooth spaces and measure the difference in the height of the displacement. The resulting error is mainly due to eccentricity, and this is the usual definition, but there can be a displacement of the measuring medium due to a local pitch or form error.

3.2.5 Sizing of the teeth

The size of a gear can be measured in a variety of ways and the following methods are the most practical and widely used:

Fig. 215. Comparative method of pitch checking
(inset) Close-up of stop and measuring feeler

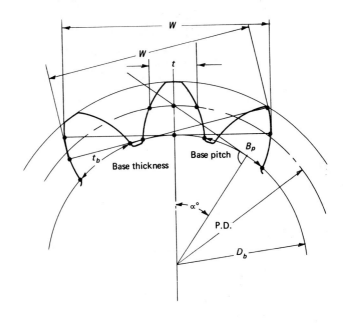

Fig. 216. Base tangent dimensions

) Base tangent

This is the best and most accurate since it can be performed with a micrometer and does not require 'feel' or operator experience to get good results. The method makes use of one of the basic properties of the involute (Fig. 216) and means that the dimension is constant irrespective of the point of tangency between the flank of the gear and the anvil of the micrometer. The base tangent dimension is found as follows.

Let span dimension over N_m teeth = W.

Then $W = (N_m - 1)$ base pitch + thickness at the base circle

$W = (N_m - 1) B_p + t_b$

To find the tooth thickness at the base circle diameter t_b.

$$t_b = D_b \left(\frac{t}{\text{P.D.}} + \text{inv } \alpha^\circ \right)$$

while the number of teeth over which the span measurement is taken can be found from:

$$N_m = 0 \cdot 5 + N \left(\frac{\alpha^\circ}{180^\circ} \right)$$

The above formulae cover the case for spur gears, and for helical gears all dimensions and calculations can be made in the transverse plane and the normal base pitch and base circle tooth thickness found by multiplying by cosine base helix angle so that the span measurement is made normal to the helix.

Special micrometers can now be purchased for this operation which have large anvils and give a precise location on the flanks of the gear.

Sizing by means of measurement over two pins placed in diametrically opposed tooth spaces is accurate although not quite as convenient as the base tangent method. The method of arriving at the tooth thickness from a given pin size is as follows (see Fig. 217):

$$M_1 = 2 (A + P_o)$$

$$\cos \varphi_1{}^\circ = \frac{r \cos \varphi^\circ}{A}$$

$$= \frac{r_b}{A}$$

$$t = 2r \left(\frac{\pi}{N} - \text{inv } \varphi^\circ + \text{inv } \varphi_1{}^\circ - \frac{P_o}{r_b} \right),$$

where N = number of teeth and t = tooth thickness.

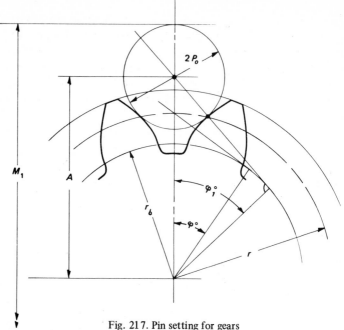

Fig. 217. Pin setting for gears

When measuring over odd numbers of teeth M_2,

$$M_2 = 2\left(A \cos \frac{180°}{2N} + P_O\right).$$

For helical gears all dimensions are in the transverse plane and

$$t = 2r\left(\frac{\pi}{N} - \text{inv } \varphi° + \text{inv } \varphi_1° - \frac{P_O}{r_b \cos \alpha}\right),$$

where α = base helix angle.

(b) *Sykes comparator*

Again this uses one of the basic principles of the involute, the constant chord (Figs. 218 and 219). This too is a comparative check since the instrument is zeroed against a master clock and the tooth thickness of the gear compared to it. The device is extremely useful

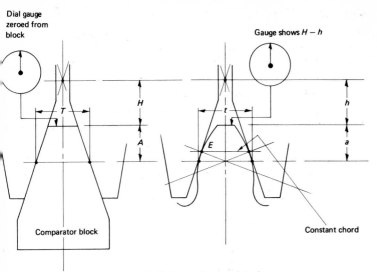

Dial gauge
zeroed from
block

Gauge shows $H - h$

H

T

t

A

E

h

a

Comparator block

Constant chord

Fig. 218. Schematic diagram of gear tooth comparator

Fig. 219. Sykes type comparator

when checking large gears since the instrument can be taken to the machine and the gear checked *in situ* without unclamping.

One of the laws of gear tooth contact in the involute system is *The common normal to the tooth curves must pass through the pitch point.* It is clear from the diagram, therefore, that the point of contact E depends only on the tooth thickness, and for a given pitch the chord formed by point E is constant irrespective of the number of teeth.

In operation the comparator jaws are set to the required tooth thickness and the dial indicator set to zero. When the instrument is placed over a gear tooth of the same thickness therefore the dial should again zero and any reading on the dial represents the difference in apex height H - h.

$$H = \frac{T}{2} \cotan \Phi° - A,$$

$$\text{and } h = \frac{t}{2} \cotan \Phi° - a$$

Since the comparator takes its reading from the addendum of the gear as well as the thickness it follows that the outside diameter of the gear must be machined first to the correct size otherwise the comparator reads the difference in addenda in addition to the difference in apex height.

(c) *Caliper setting*

This is an obvious means of measurement; it can also be carried out without removing the gear from the machine but a certain amount of 'feel' is required and it is not an accurate method owing to the lack of sensitivity.

From Fig. 220 it can be seen that for spur gears the chordal addendum A = addendum + a

where a = R versine $\alpha°$

and the chordal tooth thickness T = $2R \sin \alpha°$.

For helical gears the equivalent gear in the plane at right angles to the helix (i.e. the normal section) is found.

$$\therefore \text{equivalent P.C.D.} = \frac{\text{P.C.dia}}{\cos^2 \text{H.A.}} = 2R_1,$$

$$\text{equivalent } NT = \frac{NT}{\cos^3 \text{H.A.}}$$

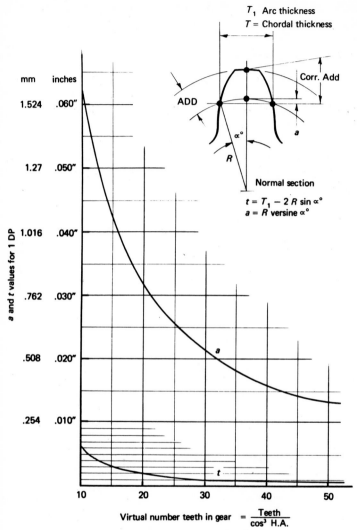

Fig. 220. Caliper settings for gear teeth

These are approximations but are accurate for all practical purposes.

$$\therefore a = R_1 \text{ versine } \alpha° \text{ and } T = 2R_1 \sin \alpha° | \text{where } \alpha\omega = \frac{T_1}{2R_1}$$

The graph shows values of a and t for a given number of teeth of unit pitch and for any other pitch the values should be divided by the normal diametrical pitch. The value t is the difference between the arc and chordal tooth thickness in the normal plane.

3.2.6 Composite checking

This is an excellent production check and enables a general standard to be laid down which can quickly and easily be assessed on the shop floor if necessary without resorting to the more sophisticated analytical checking machines. The equipment necessary to carry out the check is also more suited for use under shop conditions and consists of rolling the gear to be checked with a suitable master gear of known accuracy. This type of check can be further divided into two categories — (a) double flank testing and (b) single flank testing.

(a) *Double flank testing*

The gear to be checked is rotated in tight mesh with a master gear or

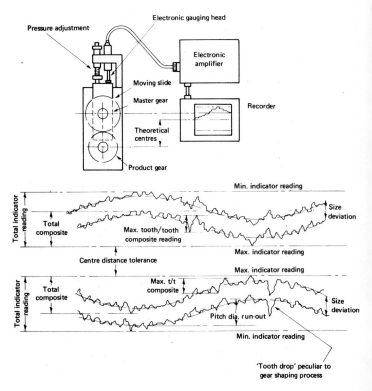

Fig. 221. Double flank roll test

worm so that contact takes place on both flanks without backlash. One member is fixed and the other spring-loaded into it to remove all backlash. As the gears rotate, the errors in engagement cause the axes to separate and this variation in centre distance is recorded on a suitable dial indicator or recorder. The conditions of engagement, of course, are never achieved in practice and consequently the results require careful analysis. However, the check is quick and simple and does not involve sophisticated checking facilities so it forms an excellent means of comparing the gear to an accepted standard. A typical graph obtained under these conditions is shown in Fig. 221. The high frequency error can be shown to have a period equal to one tooth pitch. This is known as 'tooth to tooth' error and is made up of errors in profile and adjacent pitch from two flanks. It is difficult to break these down further into individual errors but this is not necessary if the combined error is within the agreed standard. The low frequency wave form occurs in each revolution of the gear under test and is caused by eccentricity of the gear relative to its mounting. Fig. 222 shows a Goulder gear tester (which is typical of this type of equipment) in use.

(b) *Single flank testing*

Although the double flank test has been in use for many years the

Fig. 222. Goulder double flank gear tester

single flank test is a comparatively new innovation and not many instruments are commercially available. The technique has one big advantage in that it does simulate the conditions under which the gears mesh and therefore does offer a direct measure of the errors.

Some machines use interchangeable friction discs which have the same ratio as the gear under test. One is attached to the driving gear while the other is co-axial with, but independent of, the driven gear. The variation in the ratio transmitted is picked up by a sensing head and recorded on a suitable chart. The machine developed at the N.E.L., East Kilbride, uses radial gratings to sense the variation in the velocity ratio of the two gears being tested. A radial grating is mounted co-axially with each gear and the gear pair driven by a small electric motor. In order to maintain single flank contact a small restraining torque is applied to the driven shaft by a simple friction brake. The variation in angular transmission is sensed by read-out heads employing the moiré fringe optical principle. These heads generate voltage pulses which are proportional in frequency to the speed of rotation of the

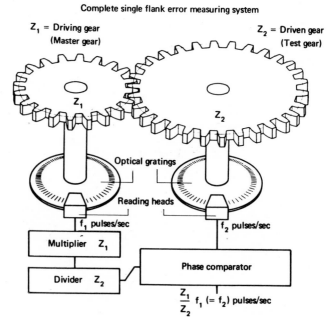

Fig. 223. Single flank testing with optical gratings

shafts. If the gear ratio were unity and the gears perfect then the two channels would have pulses of identical frequency. Errors have the effect of shifting the phase relationship of the pulse trains and if fed to a phasemeter a voltage is obtained which is proportional to the error in transmission and this enables a record to be made as the gears are rotated (Fig. 223).

The graph of Fig. 224 produced under these conditions is similar to that produced by the dual flank method but is easier to analyse since it reproduces the functional conditions. Contact between the teeth on the dual flank test can occur at two, three, or four points at the same time thus the error presented is a composite one. The error presented with single flank testing is still a composite picture but contact can only occur at either one or two points at the same time so that the com-

Single flank error charts of a pair of ground master gears
1 module; 20° pressure angle; 50 teeth. Pitch diameter ≅ 2 in (50 mm)
Full scale deflection 2 arc minutes = .0006 in (15 μ)
One division   3 arc seconds = .000015 in (0.4 μ)
Magnification 2750 X

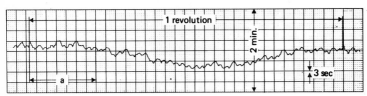

Right hand flanks
Chart speed 6 in/min (150 m/min)

Single flank error charts of a pair of hobbed gears
1.5 module; 20° pressure angle; 33 teeth. Pitch diameter ≅ 1.96 in (49.5 mm)
Full scale deflection   8 arc minutes = .0023 in (60 μ)
One division   12 arc seconds = .00006 in (1.6 μ)
Magnification 700 X

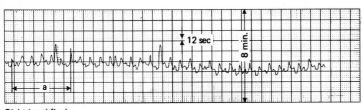

Right hand flanks
Chart speed 6 in/min (150 m/min)

Fig. 224. Typical results produced by single flank error measuring system

posite effect is not so great. This means that although both techniqu
record the profile, pitch and eccentricity error as a composite pict
the measurement is further influenced by the number of simultane
contact points. Since these contact points occur on different flar
and at varying positions it follows that the difference in local errors
the flanks further complicates the record taken. Fig. 225 show:
graphical analysis of profile errors when single flank testing — it I
been assumed that no pitch or eccentricity errors are present so that t
effect of the changing contact points can be clearly reflected in t
record taken. As shown in the diagram contact occurs between flar
a_1 and c_1 and flanks b_1 and d_1 for a certain period; this is shown
double contact.

Then for a short period only flanks b_1 and d_1 are in contact (this
single contact) before the next pair of teeth e_1 and f_1 enter in
engagement. If the profiles were perfect then simultaneous cont
between a_1 c_1 and b_1 d_1 would be possible, but in practice the flar
are not perfect and vary with each other on the same gear. Thus we ha
assumed an error on flanks a_1 b_1 and c_1 d_1 as shqwn and it follo
that if the sum of the errors on flanks a_1 c_1 and b_1 d_1 at a given inst:
were equal then both contact points would be maintained but t
driven gear would change its velocity. If the sum of the errors is n
equal, however (as shown in the diagram), only one of the cont:
points can be maintained and the flanks with the largest combin
error take control and this is reflected in the single flank record. T
diagram shows the combined errors of the mating flanks at any c
instant of contact and also the resultant graph of these combin
errors. The only true reflection, therefore, of the profile errors betwe
a pair of flanks is in the period of single contact. In practice only o
pair of flanks is in engagement at any one instant, even during t
phase of double contact; unfortunately if the errors are large t
contact could alternate between one pair of tooth flanks a_1 c_1 a
the next b_1 d_1.

Eccentricity is possibly the easiest error to identify from the grap
as if no other errors were present it would show up as a sine wa
curve of amplitude equal to the eccentricity and period equal to o
revolution of the eccentric part, while pitch errors show up as a bod
displacement of the profile curve up or down on the vertical axis.

It will readily be appreciated that with both the single and dou
flank tests the total errors recorded can be changed by altering t
phase relationship of the teeth. The errors in the two gears can

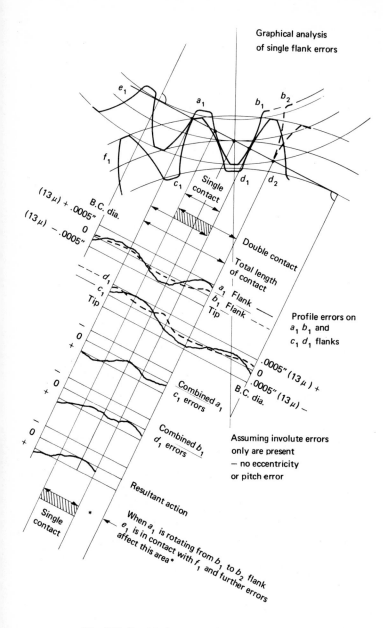

Fig. 225. Graphical analysis of single flank errors

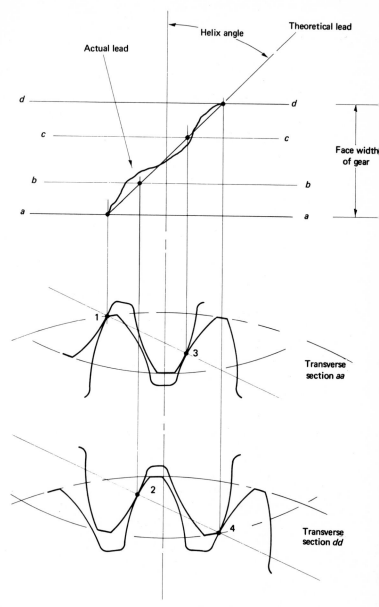

Fig. 226. Single flank checking of helical gears

added or subtracted by remeshing them in a different angular position — this has some merit when considering 1:1 ratios since the optimum meshing conditions can then be found.

So far the spur gear only has been considered — the analysis of the single flank checking of helical gears is far more complex. It has already been shown that the curve obtained by single flank checking is influenced by the number of simultaneous points in contact. Fig. 226 shows a helical gear where the helix angle and face width are such that there is complete helical overlap. On the transverse section *aa* the contact is at points 1 and 3, on section *bb* the contact is at points 2 and 4 with the same flanks in contact. The contact shifts to point 3 on section *cc* and to point 4 on section *dd* for the same pair of flanks. It can be seen, therefore, that if the profile errors only are considered then the largest total involute error between a pair of flanks at a given section through the face width determines the actual error in motion. The contact points shown are theoretical and are not maintained in practice. The position of largest error will determine the actual contact point and an analysis similar to that carried out in Fig. 225 will be necessary. This is based on the assumption that no lead error is present. However if the lead is as shown in the diagram then a further variable

Fig. 227. Goulder analytical gear checking machine

is introduced. As shown contact could only occur at the high point
lead error which is positioned on section *bb*. If the lead error w
constant at all points down the involute profile it could be said t
only the profile errors at section *bb* need be considered. Unfortunat
this is not necessarily so since what appears as a lead error could be d
to a local profile error. If large profile errors are present then they w
be recorded on the appropriate lead trace as an error. It will be appre
ated, therefore, that to analyse the single flank chart for a helical g
is virtually impossible unless separate involute and lead checks are a
taken.

3.2.7 Analytical checking machines

All the facilities for checking the various elements of a gear can
incorporated in a single machine as illustrated in Fig. 227. The use
such a machine enables a complete analysis of a gear to be made
that the dominant error and the possible cause may be determined.
shown the machine is equipped with means of checking and recordi
the errors in involute profile, lead or helix angle and automatic pit
testing. The composite dual flank test can also be performed on t
equipment by rolling with a suitable master gear.

PART 3.3: CORRECTED MACHINES

3.3.1 Methods available

On any machine tool the accuracy of the work produced is dependent on the deviation from the theoretical movement between the work and the cutting tool. These errors can be due to static and dynamic errors in the machine, errors in cutting tool and workpiece, thermal changes, friction etc. On hobbing machines in particular, however, the kinematic error is extremely important in that a gear can be generated perfectly only if the ratio between the work and the hob remains constant. Any deviation of the cutting edge from its theoretical position relative to the work at the instant a profile point is generated causes profile errors; momentary deviations in motion between the generation of the profile points are not so important.

Fig. 228 shows the effect on profile deviations of the position of the cutting edges relative to the gear flank — the broken lines indicate the

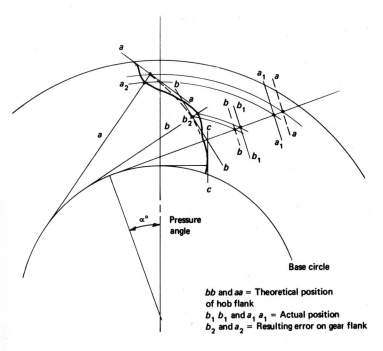

Fig. 228. Profile deviations caused by errors in position of cutting edges

correct positions and the full lines the actual positions. The gear pro
is built up of a series of flats but the flat a is out of position at a_1
thus causes the local error shown as a_2. Suitable means are available
check most elements on a machine for stiffness, static-deflection, ba
lash, error in motion of leadscrews and so on, but only in recent ye
have means been developed for checking errors in transmission.

The most important of these techniques are the following:

(a)   smoked glass technique;
(b)   friction disc;
(c)   the magnetic disc system introduced by Stepenak in Czec
        slovakia;
(d)   the seismic mass system developed by Professor Opitz
        Aachen;
(e)   the optical system using diffraction gratings developed
        N.E.L.

3.3.2 Smoked glass technique

This technique was introduced in 1929 and was mainly due to
pioneer work of the late Dr. Tomlinson. The equipment consi
basically of an accurate divide plate mounted on the end of the h
spindle or table worm shaft. This plate actuates a micro-switch and
impulses set up by the interrupted operation of the switch are record
by a diamond scribing point on a smoked plate carried on the mach
table. If the transmission is uniform the spacing of the marks on
plate should be equal, errors in motion being indicated by uneq
spacing.

3.3.3 Friction discs

This technique involves mounting friction discs on the work tai
and hob spindle and recording the variation in relative movement.

Fig. 229 shows schematically the general principle which invol
turning the hob head through $90°$ so that the axes of rotation of b
hob and work spindles are parallel to each other. One friction disc
forms part of the measuring head and is mounted on the hob spin
and this is suitably pre-loaded against a further friction disc D_t mount
on the work table. The discs are hardened and ground and mount
concentrically on their relevant axes. The diameter ratio of the di
is obviously important since any departure from the required value w
show as a progressive error on the chart. This can be overcome, ho

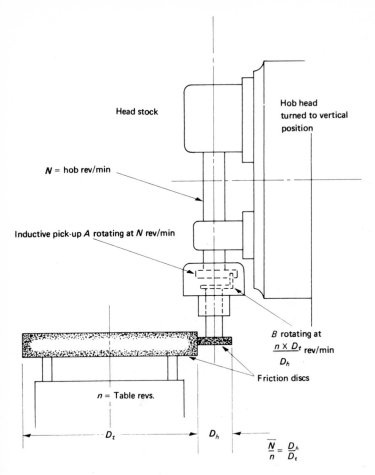

Fig. 229. Checking motion by friction discs

ever, by providing a slight taper on the periphery of one of the discs so that the velocity ratio can be adjusted slightly. The friction disc D_h and shaft B on the measuring head are driven by the table disc D_t and the difference in angular velocity of the hob spindle and the shaft B is sensed by the inductive pick-up in the recording head.

3.3.4 Stepenak system

Probably the first of the more sophisticated electronic measuring devices to be introduced in the last ten years, this method provides

V

very high accuracy in the order of 1 second of arc or less. It
the advantage that it accurately measures the total kinematic e
in the machine. Unfortunately however it is not suitable for l
frequency errors. The error of table motion relative to the hob spi
is measured by means of a magnetic scale which is generated o
magnetic drum by the hobbing machine itself. An index plate or gea
a suitable number of teeth is placed on the hob spindle and the to
frequency is detected by an inductive pick-up head. This is used a
reference frequency which is recorded on to the drum while
machine is running. During the measuring process the drum is rota
at high speed which transforms the errors in their frequency and
the use of filters the higher frequency errors are eliminated. This t
averaging method shows up the errors in the kinematic drive due to
worms and wheels but errors in other parts of the drive which have h
frequency are shown only to a limited extent.

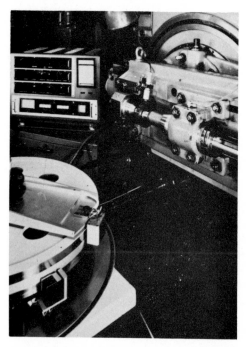

Fig. 230. Checking the accuracy of a 1 m capacity gear hobbing machine, usir
Temac electro-magnetic measuring equipment

The range of frequency is limited to a maximum of approximately four cycles per second, which limits the number of high speed components in the kinematic drive and the analaysis of tooth frequency errors.

On the other hand the system is excellent for the measurement of very large hobbing machines where the frequencies of the principal elements are extremely low (table speeds of 1/20 rev/min are fairly common).

Fig. 230 shows a typical set-up for correcting a 1 m hobbing machine; the magnetic drum on the table, the system for measuring the tooth frequency of a gear on the hob spindle and the associated electronics can be clearly seen.

3.3.5 Seismic mass device

This system was developed by Professor Opitz and his team at Aachen University in Germany and indeed is now widely used in that country by machine tool companies as an accurate means of checking the kinematic drive. This device is complementary to the Stepenak system in that it is ideal for the higher frequencies, but not for the low frequency errors where the Stepenak system comes into its own.

The pick-up unit itself consists of a seismic mass mounted on a frictionless pivot in a suitable housing. Damping is introduced at one end by means of a permanent magnet while an inductive pick-up at the other end measures the error. In this case the error is the difference in speed of the housing and the seismic mass. When fitted to a spindle then, because of its inertia, the mass will move at the mean rotational speed of the spindle to which it is attached whereas the housing will follow its non-uniform motion. One unit is mounted on each spindle to be measured thus the non-uniformity of motion of both hob spindle and table can be recorded and the difference in the signals gives the kinematic error. Any error in non-uniformity of motion caused by the initiating drive, e.g. main motor or belts, is reflected in both hob and table motions, but when comparing difference in motions it disappears.

During the work carried out at Aachen an analysis was made of the sources of the most dominant errors; this revealed that 51% of the dominant errors lay in the worm drive. This figure has possibly reduced in the last few years owing to the increasing attention paid to the accuracy of worms and wheels and the improvement in the master machines for producing them. The next largest dominant error was found to be in the index change gears, closely followed by the hob spindle drive, whereas some 23% were without dominant error.

3.3.6 Optical system

Initially developed by the N.E.L. at East Kilbride and then commercially by Goulder, the basic principle has already been described in detail on page 304. The system itself seems to offer great possibilities, particularly as the accuracy of the radial gratings can be improved. Accuracy was in the order of two seconds of arc but techniques have improved and this figure has been reduced to a half, which makes the order of accuracy for the optical system comparable to that for the other systems. Goulder's work has added to the flexibility of the technique and portable heads are being developed for various applications; typical of these is the machine shown in Fig. 231.

3.3.7 Corrector devices

In 1960 David Brown introduced a hobbing machine with automatic error correction based on the N.E.L. optical system. This sensed the error in motion and by means of a suitable servo system the table drive motion was accelerated or decelerated as required. In 1961 Pfauter of Germany also introduced an automatic error correction system using an angular displacement pick-up on the hob spindle and the work table. The relative error in motion was also compensated by means of a servo-operated device on the table worm shaft. Unfortunately the frequency limit and response characteristics of neither machine is known to the author, although the accuracy of the David Brown machine was believed to be better than four seconds of arc.

Fig. 231. Goulder single flank checking equipment

Although at that time this degree of accuracy was outstanding it has now become more commonplace and can be achieved without automatic error correction. Some companies have claimed this degree of accuracy merely by care and attention in building and by use of one of the above mentioned measuring devices during the various stages of assembly of the machine. Four seconds of arc is still no mean achievement, but it is remarkable how building accuracy improves once accurate means of measurement are available.

Churchill Gear Machines have now extended the work started by D. Brown on automatic error correction but are using the inductosyn technique in preference to the optical grating. This electronic counter ensures that the relative motions of the principal components are correct irrespective of the errors in the worms and wormwheels.

Accuracy is achieved by measuring the angles of rotation of the table and worm drive shaft, comparing them and applying a servo-controlled correction.

A 360-pole inductosyn is used to measure the table angle. This device consists of a stator which is fixed to the machine bed and a rotor which is fixed to the table. The stator produces a sine/cosine output, the cycle repeating 180 times for each revolution of the table. A resolver driven by the worm shaft is used to compare the angles of rotation of the table and the worm shaft. It is geared to make 180 revolutions for each revolution of the table.

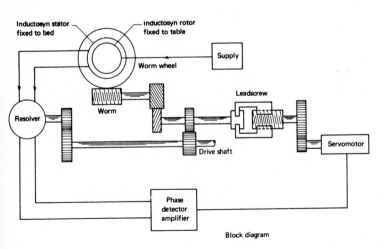

Fig. 232. Churchill gear machine electronic corrector for PH.36.15 hobber

The errors between the inductosyn and resolver give outputs whi
operate a mechanical corrector through a conventional servo syste
A servo motor working through a screw causes axial motion to be giv
to one of a pair of helical gears in the worm drive which advances
retards the angle of table rotation with respect to that of the drive sha

The system is shown diagrammatically in Fig. 232 and is designed
respond to ten cycles per second. It will correct cumulative errors
the table and any other errors within its response range.

It is possible to extend the application of the system to give a wi
range of correction, say between the hob and the table. This wou

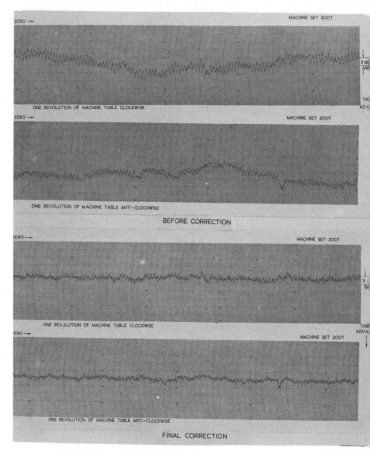

Fig. 233. Hob to table relative motion errors; final correction

make it possible to measure all 'once per revolution' errors on shafts rotating at less than 600 rev/min. Errors at flute frequency of the hob could only be recorded at fairly low speed, for example a ten flute hob at 60 rev/min.

In 1967 W. E. Sykes carried out a joint exercise with N.E.L. which involved the use of their portable grating unit as a reference to measure the errors in a 48in diameter hobbing machine. One grating was mounted on the hob arbor and the other on the work table and the relative errors in angular motion were recorded. The fundamental error in the wormwheel was found to be nine seconds of arc and a short term cyclic error from the worm was found to be three seconds of arc. A mechanical corrector drive was then fitted to the worm shaft. This involved a series of cams which were profiled to compensate the cumulative error in the wheel and the cyclic error from the worm for both directions of rotation. The accuracy of the machine was then measured again with the portable gratings and it was found that the cumulative error in the wheel was reduced to three and a half seconds of arc and the worm cyclic error to two seconds of arc, as shown in Fig. 233. A trial gear was then produced on the corrected machine to prove the result; this was a 108T 10 C.D.P. wormwheel and the resulting accuracy is shown in Fig. 234.

The Sykes designed corrector is shown in Fig. 235. Once the kinematic errors in the drive are known this device will continuously compensate for them. The corrector is mounted on the table worm shaft and consists of two sets of cams and a differential. One set of cams rotates at worm shaft speed and the other at table speed; thus through a

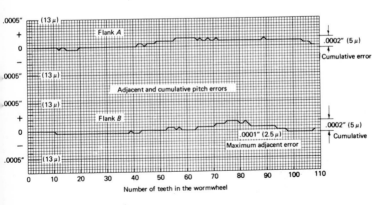

Fig. 234. 108 tooth wormwheel produced on machine after correction

Fig. 235. Automatic corrector device

differential to account for the worm and wheel ratio the worm shaft is accelerated and decelerated to correct both cyclic and cumulative errors.

SECTION 4:

CUTTING FORCES

Although the need for a high order of accuracy and stiffness in the kinematic train of gear-cutting machines has been realized by manufacturers for a number of years it has not been possible to achieve the ultimate in the design owing to lack of knowledge of the cutting forces involved. In order to improve their range of gear cutting machines and achieve a better understanding of the problems W. E. Sykes approached Cambridge University Engineering Department in 1965 and set up a programme of research into cutting forces and their effects. In three years a successful dynamometer was developed and measurement of the forces involved in hobbing, shaping and shaving was carried out. The following is a résumé of the results achieved during this period.

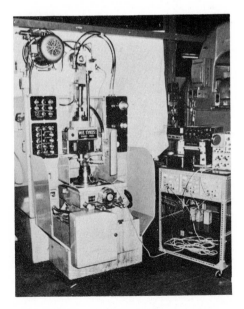

Fig. 236. Gear shaper and dynamometer at Cambridge University

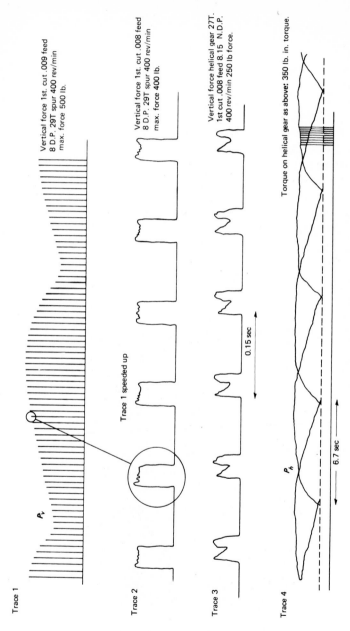

Fig. 237. Shaping forces for spur and helical gears

It should be mentioned that the work on hobbing was aided by the co-operation and assistance of Messrs. Drummond and Churchill and a grant from the Science Research Council.

PART 4.1: GEAR SHAPING FORCES

Fig. 236 shows a Sykes model V10B set up with the dynamometer, relevant electronics and recorder necessary for measuring the forces. Fig. 237 gives the results obtained on an 8 D.P. spur gear of twenty-nine teeth in a 0.4% carbon steel EN8. The trace is of the first cut of a three-cut cycle using a feed rate of 0.23mm measured around the pitch circumference and taken at 400 strokes per minute over the 25mm face width. The maximum cutting speed achieved over the face of the gear is approximately 40 metres/min; this of course varies sinusoidally due to the crank mechanism so that the speed at start and finish of the cut is considerably less. The number of teeth in simultaneous engagement varies with the tooth phasing and this accounts for the change in cutting force per stroke.

The third trace is for an equivalent helical gear of twenty-seven teeth 8.15 N.D.P. 26° H.A., also showing the vertical cutting force measured on the first cut of a three cut cycle at 0.2mm feed. The

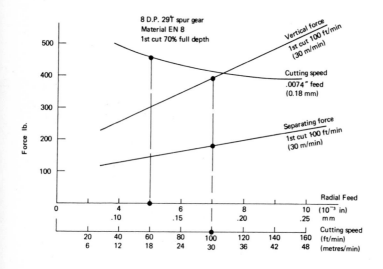

Fig. 238. Cutting forces, gear shaping

change in pattern of the force is to be expected since the helical action results in a change in the number of teeth in contact as the cutter strokes through the face width of the gear.

Fig. 238 shows the effect on the vertical cutting force when varying the feed and speed and it can be seen that the force level drops slightly

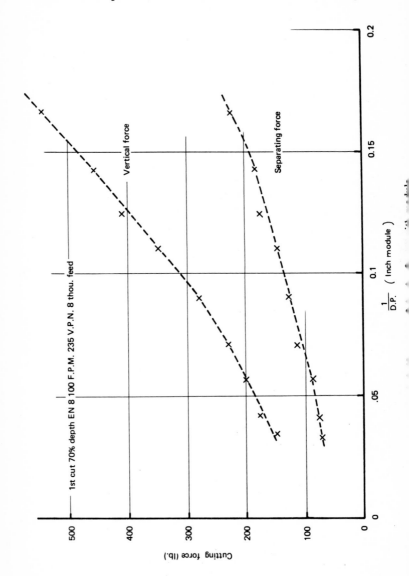

as the cutting speed is increased. The vertical and separating forces increase linearly as the feed increases and this has considerably more effect on the force levels than an increase in speed. The effect of the pitch of the gear on cutting force is shown in Fig. 239 and over the pitches tested the force level varied directly with the pitch.

PART 4.2: GEAR HOBBING FORCES

4.2.1 Spur gears

Fig. 240 shows the principal forces measured, their orientation to the tool and the work and the terminology used. Although the forces for both climb and conventional hobbing were measured, Cambridge found that the traces were smoother with climb hobbing and since this technique predominates in production hobbing the rest of the work was carried out using this method. The effect of cutting speed on the three components of force is minimal; as the speed increases the vertical and separating forces drop slightly whereas the tangential component increases to a small degree.

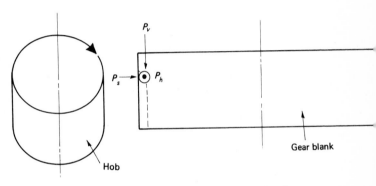

Fig. 240. Measurement of principal hobbing forces

Fig. 241 shows that the force levels rise considerably as the axial feed and/or the pitch of the gear are increased. The vertical and separating forces rise very steeply for both variables whereas the tangential component rises to a lesser degree.

It can be seen from the evidence presented in the graphs that force in the order of one ton are possible when cutting coarse pitch gear (5 D.P.) at high feed rates.

Fig. 242 compares the forces measured for both climb and conventional hobbing, and while the component of force tangential to the tool is slightly larger with conventional hobbing compared to climb hobbing the resultant force is some 25% larger and acts in a different direction. This could account for the difference in behaviour of some machine when using both techniques of cutting.

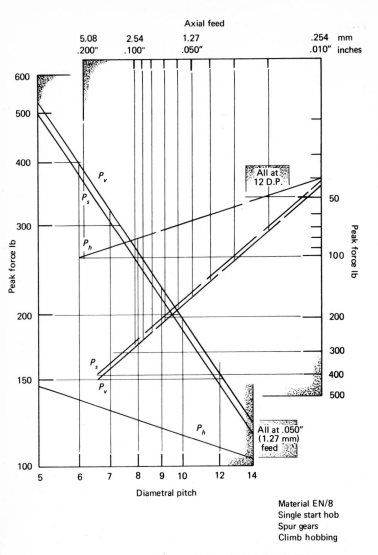

Fig. 241. Peak cutting forces *v.* D.P. and axial feed hobbing

4.2.2 Helical gears

The work on spur gears was carried out with a conventional hobber whereas the work on helical gears was performed on a Drummond

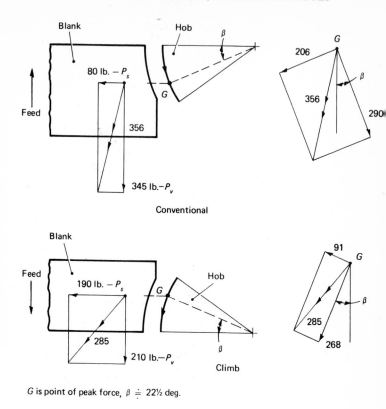

Fig. 242. Comparison of forces in climb and conventional hobbing

oblique machine (Fig. 243), whereby the hob head is inclined at the helix angle of the gear and is fed down the pitch plane. The feed rates stated on the following graphs are measured down the helix and are not therefore directly comparable with the feed rate used on a conventional hobber with differential. The feed rate is measured down the axis of the gear and is greater than the feed down the helix by the secant of the helix angle.

The graph in Fig. 244 (page 330) shows the peak forces registered for all three components of force when climb hobbing gears from 0° to 45° helix angle, and Fig. 245 shows typical force traces. All gears were cut at a constant feed rate of 1.36mm down the helix and it can be seen that the force levels increase drastically with helix angle; for example the

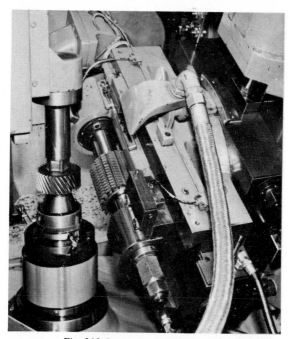

Fig. 243. Drummond oblique hobber

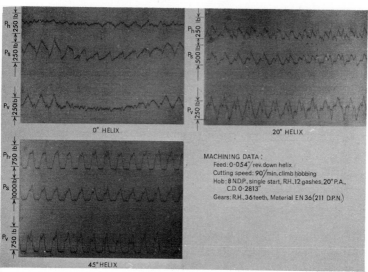

MACHINING DATA:
Feed: 0·054″/rev. down helix
Cutting speed: 90′/min. climb hobbing
Hob: 8 N.D.P., single start, R.H., 12 gashes, 20° P.A.,
C.D. 0·2813″
Gears: R.H., 36 teeth, Material EN 36 (211 D.P.N.)

Fig. 245. Traces of cutting forces for helix angles of 0°, 20° and 45°

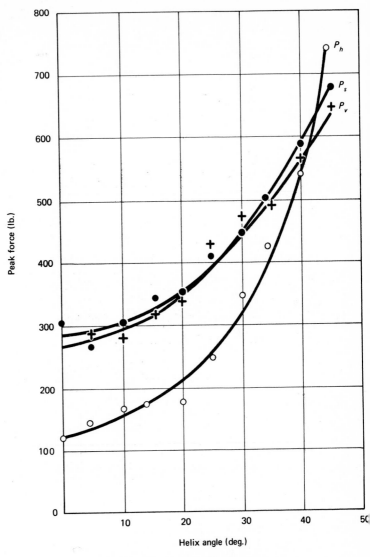

Fig. 244. Peak cutting forces hobbing helical gears

separating and vertical forces double between 10° and 40°. It must be
appreciated, however, that these results were obtained from a fixed
number of teeth (thirty-six) in the transverse plane which means that the

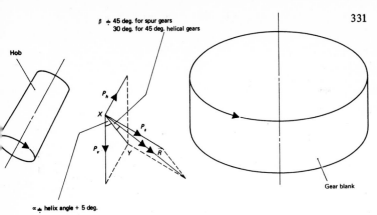

Fig. 246. Determination of resultant force and direction

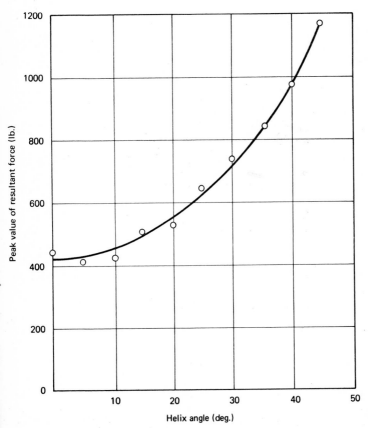

Fig. 247. Peak resultant forces plotted against helix angle

virtual numbers of teeth, i.e. in the plane normal to the helix, ⁣
37.7 and 80 respectively and the feed down the gear axis will h⣿
increased by the secant of the helix angle. The hob therefore is doin
great deal more work.

Fig. 246 shows how the resultant force can be determined from ⁣
three forces measured and the direction in which it acts while the gra
in Fig. 247 shows the peak resultant forces plotted against helix ang
One obvious comment is that the peak forces will increase as the h⣿
wears, so it must be appreciated that the forces shown in the grai
were recorded with newly sharpened tools and the values shown co⣿
increase by some 50% for 0.13mm wear on the hob.

PART 4.3: GEAR SHAVING FORCES

This work has proved to be quite complex and the investigations are still going on at Cambridge so that it is difficult to draw any definite conclusions at this time. The first stage was to attempt to measure the three components of cutting force and Cambridge decided to mount the gear between two identical dynamometers which were similar in principle to those used so successfully in the other work. The set-up on a Sykes type gear shaving machine is shown in Fig. 248.

Fig. 248. Measuring shaving forces on a shaving machine at Cambridge University

At the time of writing the work was in the preliminary stages and it is not proposed to show the traces obtained since the results were not conclusive.

Carbide cutting

Sintered carbide has been used for many years for the machining
steel, but until recently very little success had been achieved with t
gear generating processes such as hobbing or shaping which do n
produce a chip of constant size. It had been used successfully in t
production of non-ferrous gears but the hobs were usually carbi
tipped or inserted blade and were not prone to chipping as would
the case if used on steel.

A major breakthrough was achieved in 1968. The new hob produc
by Mikron after many years of research was made from solid carbi
and in a new grade which gave much better results. Once a hob h
been designed specifically for use with carbide tools. They involve
further and in 1969 a number of companies developed machines whi
they claimed were capable of utilizing carbide tools.

After the first successful tests the development took place in thr
stages. The first stage involved standard machines and it was found th
the hob had much greater potential than the machine. The second sta
involved a 'beefed up' version of a standard machine with heavier h
head, larger motors, plus the facility for higher hob speed. The thi
stage machines are now just starting to make their appearance and ha
been designed specifically for use with carbide tools. They involve
structure much stiffer than previously conceived, new and stiff
kinematic trains and high speed potential up to some 2000 rev/min
the hob. Special attention has been paid to bearing stiffness and t
dissipation of heat in the machine itself apart from the removal of t
hot swarf from the working area of the machine. The hot swarf, if n
removed, causes local deformation of the machine structure and spec
care in design is necessary to ensure that any thermal expansion of t
machine occurs uniformly.

The performance of the stage 3 machines is difficult to predict
this time since very little data have yet been published and the on
results available are those from the first two types. To generali

334

regarding speeds/feeds and productivity is always dangerous, but it would appear reasonable to expect the carbide hob to run at approximately three times the surface speed of an equivalent high speed steel hob and the tool life to be comparable.

Experiments to date seem to indicate that the carbide hob requires fairly heavy cuts to avoid chipping, and feed rates can also be increased. Cutting speeds between 750 ft/min (240 metres/min) and 1000 ft/min (300 metres/min) have been used successfully together with feed rates of the order of 0.200 in (5 mm) per revolution. Owing to this apparent requirement for heavy cuts the indications are at the moment that finish cutting with carbide may not be desirable and that secondary finishing techniques such as shaving and finish cold rolling may still be necessary.

The economics of carbide hobbing are still difficult to assess since the full potential of the hob has not yet been realized. The initial tests were carried out on machines designed for maximum hob speeds in the region of 500 rev/min and updated to 1000 rev/min and on this basis a considerable reduction in production time has been achieved. Whether the hob is capable of being used at even higher surface speeds remains to be seen and cannot be determined until the next generation of machines running at say 2000 rev/min of the hob are available and have been evaluated. Certainly this new generation of machines poses new problems for the designer and in order to obtain the maximum dynamic stiffness a minimum number of working parts of short length and high stiffness are necessary. It appears highly likely that the main drive will be direct on to the hob arbor and in view of the high power requirements in a compact area the drive will probably take the form of a hydraulic motor. Because of the high stiffness and speed requirements the use of hydrostatic bearings would also appear to be a desirable feature. The power requirements are two to three times greater than with a conventional machine and motors of the order of 18-20 hp are likely to be used.

Most cutting is carried out dry, although coolant can be circulated through the machine structure and over the work table area in order to dissipate the heat generated by the cutting action and the hot swarf. Coolant could be used providing it could be guaranteed that it would be in contact with the tool and the work at the point of cut at all times. If dry spots occur small cracks are likely to be propagated on the cutting edges, leading to early failure.

The performance of the hob is increased considerably by rounding off the edges of the teeth slightly. Mikron advise that this rounding should be of the order of 0.03 mm (0.0012 in) − 0.06 mm (0.0024 in) It is achieved by a tumbling technique, the hob being placed in vibrator with a suitable abrasive medium and the desired radius achieve in approximately 60-80 min.

Hob sharpening does not appear to be a problem and certain man facturers have already claimed that existing machines can be convert for this purpose while others have introduced new machines. In ord to avoid the propagation of cracks during grinding it is essential to kee the heat generated to a minimum and the use of a copious flow coolant on the flute face is recommended. Diamond wheels are use operating at approximately half the speed normally used for conve tional high speed steel hobs, i.e. 6000 ft/min (200 metres), and fir cuts should be taken at slow table speeds.

A major breakthrough in gear hobbing techniques has certainly bee achieved. It remains to be seen how far the new generation of stiffe high speed machines will allow the carbide hob to be used. New grade of carbide will undoubtedly be developed now that stiffer machines a becoming available to extend their performance. Manufacturing techr ques for the hob are also under review and if the demand increases th cost should drop; both factors will considerably affect the economi of the process.

Certainly if it continues to develop we shall see typical car gea being produced in 10-20 seconds. This in turn will lead to furthe development of high speed automatic loading devices in order to ensur optimum machine utilization.

The need for extreme stiffness has been emphasized several time and for this reason it should not be assumed that all gears lend then selves to production by this technique. The work holding fixture i terms of stiffness should be regarded as part of the machine and if th nature of the component is such that slim or non-stiff fixtures ar required because of physical limitations or fouling points, optimur results cannot be expected.

High speed gear shaping

Until recently the reciprocating speeds of the gear shaping machine were limited to some 800 rev/min; with the advent of new and stiffe machines these speeds have increased to 2000-2500 rev/min. This i itself has brought no increase in productivity on gears of 1½ in (37 mm face width or larger since the cutter has not yet been developed whicl will stand the increased surface speeds.

On clutches or gears of narrow face width there is an advantage in that owing to the previously limited stroking speed the surface or cutting speed was perhaps only of the order of 60-80 ft/min (20-26 metres), whereas the material was possibly capable of being machined at 100-120 ft/min (32-40 metres) with conventional high speed steel cutters. The higher reciprocating speeds therefore give a production saving on the smaller face width gears or gears produced in the softer materials, i.e. aluminium, brass, etc.

Work is being done on the development of new cutter materials — mainly carbide — which, like hobbing, would possibly boost the cutting speed from the conventional 100 ft/min (32 metres) to say 250 ft/min (80 metres), or even 750 ft/min (240 metres). If such materials become a commercial proposition then of course the high speed potential of the new sophisticated high speed gear shapers will be realized. There is some disagreement among the acknowledged experts as to whether hydrostatic guide units or bearings are justified at speeds up to say 180 ft/min (60 metres). If speeds of 750 ft/min (240 metres) are eventually achieved, then almost certainly hydrostatic bearings will be necessary, since it is unlikely that plain bearings will be capable of standing such rubbing speeds at the close running clearance required for this type of application.

Single die gear rolling

The most recent development in the cold rolling field (and the position appears to change month by month) has been the introduction by Carl Hurth in Germany of a single rotary die machine. By using one die and not two the problem of synchronizing the speed of the dies is resolved and the heavy pressures inherent in rolling are relieved to a certain extent by a special configuration of the die itself.

The die is serrated in a similar manner to a shaving tool, thus reducing the area in contact and the pressure required. The serrations are applied in a differential or staggered pattern so that there is a suitable overlap on consecutive teeth.

The design and technology of gear cutting machines have advanced very rapidly in the last two years and there are signs of major breakthroughs on several fronts. It is very difficult to foresee at this stage which developments will be of lasting value.

REFERENCES

Barber Colman Co., 'Multi-start hobs'

J.D. Smith and D.B. Welbourn, 'Optimum infeed rates in shaping spur gears', *Machinery, 110* (June 28th 1967)

D.A.D. Cooke and D.B. Welbourn, 'Forces in gear hobbing — spur gears', *Machinery, 111* (September 6th 1967)

D.A.D. Cooke and D.B. Welbourn, 'Forces in gear hobbing — helical gears', *Machinery, 112* (February 7th 1968)

INDEX